James Smyth

The effect of the nitrous vapour in preventing and destroying

contagion

With an introduction respecting the natur

James Smyth

The effect of the nitrous vapour in preventing and destroying contagion
With an introduction respecting the natur

ISBN/EAN: 9783337126261

Printed in Europe, USA, Canada, Australia, Japan

Cover: Foto ©berggeist007 / pixelio.de

More available books at **www.hansebooks.com**

OF THE

NITROUS FUMIGATION.

Je ne connois pas de theorie qui puiſſe decider une queſtion qui intereſſe la vie des hommes ; il ni a que des expériences faites avec foin, et multipliées à l'infini, qui doivent
fervir de loi en medicine.

THE

EFFECT

OF THE

NITROUS VAPOUR,

IN

PREVENTING AND DESTROYING

CONTAGION;

ASCERTAINED, FROM A VARIETY OF TRIALS,

MADE CHIEFLY BY

SURGEONS OF HIS MAJESTY's NAVY,

IN PRISONS, HOSPITALS, AND ON BOARD OF SHIPS:

WITH AN

INTRODUCTION

Respecting the Nature of the Contagion,

WHICH GIVES RISE TO THE

JAIL OR HOSPITAL FEVER;

AND

THE VARIOUS METHODS FORMERLY EMPLOYED TO
PREVENT OR DESTROY THIS.

BY

JAMES CARMICHAEL SMYTH, M.D. F.R.S.

FELLOW OF THE ROYAL COLLEGE OF PHYSICIANS,
AND PHYSICIAN EXTRAORDINARY TO HIS MAJESTY.

PHILADELPHIA:

PRINTED BY BUDD AND BARTRAM,

FOR THOMAS DOBSON, AT THE STONE HOUSE,
Nº 41, SOUTH SECOND STREET.—1799.

[v]

TO

THE RIGHT HONOURABLE

EARL SPENCER,

&c. &c. &c.

My Lord,

As most of the following trials with the nitrous vapour were made in consequence of an order issued by your Lordship, and the other Lords Commissioners of the Admiralty, for employing it in the navy; it must afford you some satisfaction to observe, that success has followed your Lordship's decision on this, as on more important occasions: nor can you be surprised that I should be desirous of prefixing your name to a publication, which owes its existence, in great mea-

sure

fure to yourfelf. I have now no occafion
to folicit your Lordfhip's recommendation of
a meafure, which a conviction of its utility
has already led you to adopt; but I have to
thank you, which I do moft fincerely, for
the attention you have given to the fubject,
and for your candour and politenefs to the
Author..

I have the honour to be,

With the moft perfect confideration and refpect,

My Lord,

Your Lordfhip's moft obedient,

And much obliged fervant,

JA' CARMICHAEL SMYTH.

Earl Spencer.

PREFACE.

PREFACE.

THREE years have now almoſt elapſed ſince I publiſhed an account of the experiment made with the nitrous fumigation on board the Union hoſpital ſhip, and of the ſimilar trials made at the ſame time on board ſome ſhips of the Ruſſian ſquadron.

The accurate and candid narrative of this buſineſs, given by thoſe gentlemen, who undertook the conducting it, proved in the cleareſt and moſt unequivocal manner, to every unbiaſſed mind, not only the power of the nitrous vapour in deſtroying contagion, but the ſafety with which it might be employed : ſuch, however, is the force of prejudice, as to render doubtful even the evidence of our ſenſes. A conſiderable increaſe in the number of deaths, amongſt the Ruſſians, happening

pening in the month of January, a rumour was immediately propagated that this mortality was occafioned by the nitrous fumigation, which, though it might feem harmlefs in the beginning, poffeffed fome latent deleterious quality that in the end proved fatal.

Hearing of this by accident, and knowing how difficult it is to remove impreffions that have once laid hold of the public mind, I made application to Lord Spencer, and to the other Lords Commiffioners of the Admiralty, requefting that they would have the goodnefs to appoint proper perfons to inveftigate this matter fully, and to report to their Lordfhips the refult of their inquiry. My requeft was complied with, and the report of thofe gentlemen proved in the moft fatisfactory manner, that the rumour was a direct mifreprefentation of the fact; that the mortality amongft the Ruffians was owing to different caufes unconnected with the fumigation; that thofe who had been longeft

expofed

expofed to it enjoyed the beft health ; and that not a fingle inftance could be adduced of any bad confequence which could fairly be imputed to the nitrous vapour, during the three months it had been employed.

This objection being removed, another ftill prefented itfelf. It had been faid, and faid with truth, that the vitriolic acid, in a concentrated ftate (commonly called oil of vitriol) was a dangerous article to be taken on board of fhip, as there had been inftances of fhips fet on fire by it ; and that this accident actually happened to two tranfports of Admiral Chriftian's fquadron. The fact could not be denied ; but the fame objection applied, and with ftill greater force, to the ufe of fire, of gun-powder, and of ardent fpirits ; all of which are, without doubt, extremely hazardous in the hands of ignorance or of rafhnefs ; but which, when their effects are known, with the proper means of obviating them, may be employed with as much fafety as air or water.

B

To

To do away, however, every poffible objection on this head, I had proper cafes made,
one for the mineral acids, and another for the
purified nitre, with the inftruments neceffary
for the fumigation. The mineral acids were
firft put into ftrong glafs bottles, fitted with
ground glafs ftoppers, fecured with wire;
thefe were placed in a cafe lined with copper
covered with an amalgama of tin and lead,
with divifions of the fame; the interftices of
the divifions were afterwards filled up with
faw-duft: by this means the bottles were fecured from breaking, and even if that fhould
happen, the acids could not efcape, nor affect
the lining of the cafe; fo that every danger
which could poffibly arife was completely
forefeen and prevented. A cafe fimilar to the
former, but entirely of wood, with divifions
of the fame, was made for holding the nitre:
the nitre was put into four ftone or earthen
jars, adapted to four divifions in the cafe,
which had a fifth divifion alfo, the whole
width

width of the cafe, for containing the pipkins, cups, meafures, and other inftruments necef- fary for the fumigation. Thefe cafes I fent to the Admiralty for the infpection of their Lordfhips, and as they met with their appro- bation, their Lordfhips ordered fimilar cafes to be made and fent on board every fhip in his Majefty's fervice; and that the materials neceffary for the nitrous fumigation fhould be fent alfo to the different naval hofpitals and prifons. I need hardly add, that the greater part of the experiments and trials which I have now the honour to lay before the public, are the refult of this order of the Lords Commiffioners of the Admiralty.

Many of thofe communications have been fent to me, as will prefently appear, by the Board for fick and wounded feamen; but for feveral, by far the moft important, I am in- debted to the friendfhip of Dr. Johnfton, one of the Commiffioners of that Board, a gentleman whofe humanity and active zeal in

the

the fervice of his country are well known, and whofe character ftands too high in the public opinion to receive any addition from my praife. He was fortunately one of the extra-Commiffioners at Portfmouth, when fome of the firft trials were made with the nitrous fumigation, and from the happy effects produced by it, became a warm advocate in its behalf; a circumftance which firft introduced me to the honour of his acquaintance.

As for the gentlemen themfelves who have made the trials with the nitrous fumigation, I fhall only obferve, that they cannot be fufpected, in the reports or opinions they have delivered on this fubject, to have been influenced by any partiality to me; as, excepting Captain Lane of the navy, I had not the honour of the flighteft acquaintance with any of them; and feveral, though they had heard of the experiment on board the Union, had never read a word I had written on the fubject. It appears, however, very evident from

their

their letters, that there are amongft them men of obfervation and experience, and if we may judge from the important ftations in which fome of them are placed, they are men of high and refpectable characters in the fervice. Refpecting their obfervations, I may fairly fay, that in general, they carry with them the internal evidence of truth. But I fhall examine this fubject more particularly, after having prefented the public with the letters and communications of the authors.

Of the Introduction I have only to remark, that it contains thofe general obfervations on the jail and other putrid contagions, with the ufual means of obviating thefe, formerly publifhed in the Treatife on the Fever at Winchefter; and now republifhed, from a wifh to make them more generally known, efpecially to thofe perfons who are likely to be placed in fimilar fituations, or engaged in fimilar enterprifes. It does not fall to the lot of every furgeon, or even phyfician, to the

navy

navy or army, to have turned his attention to this fubject in the manner I have done, nor to have had the fame means or opportunity of information. The chief object of my life has been to render my profeffion, and the exercife of it as beneficial as poffible to mankind; my endeavours, however, to bring to perfection, and into general ufe, the prefent difcovery, have been chiefly animated, and I am not afhamed to confefs it, by the defire of rendering an important fervice to my country. She, as the great maritime power in Europe, is the moft likely to derive, and I flatter myfelf, will derive the moft effential benefit from my labours.

Chari Parentes, Chari Liberi, Propinqui Familiares: fed omnes omnium Charitates, Patria una complexa eft.

CONTENTS.

INTRO-

The READER *is requested to examine these* TABLES *with Attention, as they afford, perhaps, the most complete Evidence, of a Medical Fact, that ever was presented to the Public.*

Office for sick and wounded Seamen, &c.

A Weekly progressive State of the Sickness and Mortality among the Spanish Prisoners, confined in the King's House, at Winchester, from the first Appearance of the Jail Distemper, until the 8th of July, 1780.

Date of Weekly Accounts.	Number of Spanish Prisoners.		
	In Custody.	Sick.	Dead.
March 26, 1780	1247	60	1
April - 2, ——	1243	106	4
—— 9, ——	1475	150	10
—— 16, ——	1457	172	18
—— 23, ——	1433	142	21
—— 30, ——	1412	171	21
May - 7, ——	1388	191	25
—— 14, ——	1351	197	27
—— 21, ——	1523	205	30
—— 28, ——	1494	226	31
*June - 3, ——	1461	262	33
—— 10, ——	1437	212	26
—— 17, ——	1426	173	9
—— 24, ——	1420	167	5
July - 1, ——	1414	143	5
—— 8, ——	1433	122	2

* The time of Dr. Carmichael Smyth's going to Winchester.

A Weekly Return of the Patients in Forton Hospital, &c. from the 16th of October to the 26th of December, 1796. By D. Paterson, Esq. Surgeon to the Hospital.

	Before the Nitrous Vapour was used.				After the Nitrous Vapour was used.		
Weeks	Highest Number in the Hospital	Number discharged	Number dead	Weeks	Highest Number in the Hospital	Number discharged	Number dead
1	223	2	8	1	340	27	6
2	372	4	21	2	332	7	5
3	371	0	13	3	742	11	8
4	369	1	9	4	340	8	4
				5	456	12	1
				6	532	63	5
		7	51			128	29

A Return of the Attendants on the Hospital, or Persons belonging to the Ship's Company of the Union, who were attacked with the Contagious Fever, from the 3d of September, 1795, to the 10th of February, 1796.
(Signed) A. BASSAN, Surgeon of the Ship.

	Before the Nitrous Vapour was used.			
Quality	When seized	Recovered		Dead
Nurse	Sept. 6	————		
	7	————		
	9	————		
	11	————		
St. th Mate	15	————		Sep. 24
Nurse	18	————		
Helper	20	————		
	22	————		
Nurse	24	————		Sep. 28
Marine	29	————		Oct. 1
Washer-wo.	Oct. 6	————		15
St. th Mate	6	————		
Nurse	8	————		11
Ab.	8	————		
Nurse	14	————		
2d St. Mate	22	————		
Ab.	22	————		
Marine	22	————		
Nurse	Nov. 4	————		
Washer-wo.	4	————		
Marine	4	————		
	10	————		
Ab.	10	————		
	17	————		
	17	————		
Marines	18	————		
St. Marines	18	————		
Ab.	20	————		Dec. 4
Marine	20	————		Nov. 24
Washer-wo.	24	————		T. uncertain
	Total 30	23		8

	After the Nitrous Vapour was used.		
Quality	When seized	Recovered	Dead
Nurse	Dec. 25	Recov. in a few days	
Marine	26		Jan. 6
	Total 2	1	1

N. B. On the 15th of November the ship was fumigated for the first time, and the fumigation repeated twice a day till the 13th of December; from that time to the 26th of December, only once; but from the 16th of December to the 10th of February, twice a day, as at first.

A Monthly and Daily Return of the Convicts attacked with the Jail-Fever on board the Hulks, and received into the Sincerity Hospital Ship in Langstone Harbour, from the 6th July to the 26th Dec. 1798. By S. Hill, Surgeon to the Hospital.

Months	Days	Numbr	Month	Days	Numbr	Month	Days	Numbr	Month	Days	Numbr
July	——	16	Oct. 16	3	Nov.	1	1	Dec.	1	3
Aug.	——	66		17	1		2	1		2	3
Sept.	——	120		18	1		3	··		3	3
Oct.	1	3		19	1		4	1		4	5
	2	3		20	3		5	··		5	5
	3	7		21	4		6	3		6	5
	4	11		22	3		7	1		7	4
	5	5		23	1		8	1		8	3
	6	7		24	3		9	——		9	4
	7	3		25	2		10	7		10	2
	8	4		26	1		11	7		11	3
	9	2		27	2		12	7		12	2
	10	2		28	1		13	1		13	1
	11	6		29	1		14	··		14	2
	12	8		30	0		15	··		15	1
	13	3		31	1		16	——		16	3
	14	5					17	——		17	2
*15	9					18	——		18	2
							19	——		19 H	6
							20	——		20	2
							21 H			21	3
							22	——		22	3
							23	1		23	2
							24	7		24	1
							25	+		25	1
							*26	2		26	2
							27	2			
							28	2			
							30 H	3			
	To the 16th	77	To 31st	30			50			64

* On the evening of the 15th of October we began to fumigate the hulks, and continued to do so every day to the 10th of November, (the 9th and 10th of this month excepted;) on the 12th, the fumigation was discontinued, but resumed on the 16th, and continued without interruption to the 13th of January, 1799, though on the 26th of December the sickness had entirely ceased.

[H] On the 21st of November, eight persons ill with the jail-fever, or dysentery, were received into the hospital, from on board the Hillsborough Botany-bay ship, outward-bound; eleven more were received on the 30th of the same month, and five on the 19th of December; in all, twenty-four.

INTRODUCTION.

Respecting the Nature of the Contagion, which gives rise to the Jail or Hospital Fever.

THAT we may be able to form a more accurate judgment of the nature of the contagion, which gives rise to this fpecies of fever, we fhall examine it in four different points of view.

In the firft place, how it is generated; 2dly, in what manner it is propagated, with the circum-ftances more or lefs favourable to its communica-tion; 3dly, its effects on the human body; and, 4thly, the means of weakening its virulence, or of entirely deftroying it.

Whoever has confidered contagious fevers with that attention which the importance of the fubject demands, muft have obferved, that they are of two very diftinct kinds or *classes. The firft may pro-perly enough be called *fpecific contagions*, as they do not arife from any general quality, or procefs of nature, with which we are acquainted; and, as they have a peculiar origin, they excite difeafes of a peculiar kind; differing in many refpects from every other, but in nothing more remarkably, than in this, that the peculiar difeafe can only take place once in any individual: and there are fome perfons,

C   in

* This diftinction feems fo extremely obvious, that we fhould hardly fuppofe it could efcape the obfervation of any one who had at all reflected on the fubject, and yet I do not know any medical writer who has made it.

in whom this contagion never can produce any morbid fymptom. The number of thofe peculiar or fpecific poifons is not yet afcertained; but the fmall-pox and meafles are evidently fuch to man, and there are others peculiar to certain animals.

The fecond clafs of contagious fevers, may be named *general contagions*, as they arife from a general caufe; or they may be named *putrid*, as they will be found, in every inftance, to be the refult of putrefaction, one of thofe general fermentative proceffes, to which water as well as all vegetable and animal fubftances, under certain circumftances, are liable. That the contagion, or miafma, of the jail or hofpital fever derives its origin from this fource, admits of every fpecies of evidence which a matter of fact and of obfervation can do.

It is well known, that this difeafe is conftantly produced where a number of people are fhut up together in a clofe place, without the greateft attention to cleanlinefs, and a renewal of air. That all the excretions of the human body have made a certain advance or progrefs towards putridity, and when placed in circumftances favourable to putrefaction, foon become highly putrid. That of all the human excretions, none is more highly animalifed, or more fufceptible of becoming putrid, than the perfpiration or vapour iffuing from the furface of the body and lungs. That even the perfpiration of vegetables, confined under fimilar circumftances, becomes putrid, and in a high degree noxious to man: *a fortiori* then, we may conclude, that animal perfpiration will more quickly undergo a fimilar alteration, and muft prove ftill more noxious.

We obferve, that the contagion, refulting from animal perfpiration, fhews its baneful effects more fuddenly, and more forcibly, in proportion to its quantity,

quantity, and to its being placed in circumſtances the moſt favourable to putrefaction; conſequently, in proportion to the ſize and cloſeneſs of the place, the temperature* and moiſture of the air, and the additional or acceſſory putrid matter with which it is combined.

We obſerve, likewiſe, that the formation of this contagion is prevented by cauſes that renew the air, and carry off the perſpiration, or prevent its tendency to putrefaction.

And, it may be mentioned as an analogical proof, that a contagious vapour, differing only in degree of virulence from the human miaſmata, is conſtantly produced from water alone, or from water mixed with vegetable and animal matters, when expoſed in ſufficient quantity and under circumſtances favourable to putridity; but the ſeptic nature of the jail contagion will be farther illuſtrated, by what we have to notice of its effects on the human body,

---

* It has been alledged in objection to this, that the jail fever is more frequent in our priſons in winter than in ſummer. The fact I can neither confirm nor deny, but admitting it in its fulleſt extent, it by no means invalidates the opinion I have endeavoured to eſtabliſh. The cold of our winters is ſeldom ſo ſevere, at leaſt for any length of time, as to freeze the moiſture of the atmoſphere, eſpecially within-doors, and therefore, I run no riſk of contradiction, when I affirm, that in general, the moiſture or water diffuſed in the atmoſphere, is much greater during the winter, than during the ſummer months; but the cold, though not in general ſufficient to dry the air, is ſufficient for thoſe confined in the cells of priſons to endeavour to exclude it, although by excluding it, they muſt prevent the renewal of the air, and breathe more their own baneful atmoſphere. Cold has alſo the effect of making people in their ſituation, leſs attentive to cleanlineſs, and of inducing lazineſs and debility; all of which circumſtances, when taken into the account, will be found greatly to counterbalance the difference of temperature.

body, and of the methods of deſtroying it, or of rendering it harmleſs.

Of the Manner in which Contagion is communicated.

EVERY perſon knows that contagious fevers, whether *ſpecific* or *putrid*, are uſually propagated by an immediate communication with the ſick, either by contact or contiguity. How far the contagious atmoſphere extends, is impoſſible to aſcertain, as this muſt admit of *conſiderable* latitude*, according to the virulence of the diſeaſe, ſituation of the ſick, ſeaſon of the year, ſtate of the atmoſphere, &c. My ingenious and reſpectable friend Dr. Haygarth is of opinion, and indeed has ſhewn, that in the ſmall-pox it is much more limited than was apprehended. But it is not only from a direct communication with the ſick that contagious fevers are propagated; unfortunately, the perſons and clothes of thoſe who remain long in a contagious atmoſphere, and the excretions of the ſick, are capable (even when conveyed to a great diſtance, or preſerved for a length of time) of producing the ſame miſchief as an immediate communication with the ſick themſelves. Of this fact the examples are ſo numerous as to put the matter beyond the poſſibility of a doubt. Here again, the opinion of my friend Dr. Haygarth differs from the opinions formerly entertained by phyſicians. For, though he admits, that the variolous matter, and the more ſenſible excretions of the ſick, are capable of communicating the diſeaſe, and, if cloſe ſhut up, of retaining that power for a long time, he does not think, that the contagious vapour, immediately
arising

* This ſubject the reader will find treated more fully in two Letters addreſſed to Dr. Percival, of Mancheſter.

arifing from the fick, can be retained by the clothes
of thofe confined in the variolous atmofphere, or
by the furniture in the chambers of the fick, fo as
to communicate the difeafe to fuch as have not
themfelves been immediately expofed to it. No
one can have a greater refpeft for the opinions and
obfervations of Dr. Haygarth than I have, as no
perfon is better acquainted with his candour and
accuracy. I readily agree with him, that the dread
of thofe terrible difeafes, and the natural fears of
men, have poffibly magnified the danger beyond
reality ; that the rifk of propagating the contagion
in this manner is by no means fo great as had been
fuppofed ; and that phyficians, or even apothecaries,
are feldom fo long expofed to this atmofphere, as
to be in great danger of conveying the contagion
elfewhere ; but I cannot go fo far as to believe that
the perfons, and efpecially the clothes of nurfes
or affiftants, who are conftantly confined in the
chambers of the fick, fometimes not very well ven-
tilated, will not imbibe the contagious vapour to
fuch a degree, as to be capable of communicating
it, efpecially where they have a direft or immediate
intercourfe with a perfon fufceptible of the difeafe.
But, putting the fmall-pox and other *fpecific conta-*
gions out of the queftion, that the jail diftemper
and other *putrid contagions* are frequently conveyed
in this manner, cannot be denied. Indeed, wherever
a vapour can be diftinguifhed by the fmell, we have
the demonftration of our fenfes for what a length
of time, not only clothes, but furniture, and even
the boards and walls of houfes will retain it : therefore, in refpeft to the contagion of the jail or hofpital fever, we may fafely affirm, that it affefts not
only thofe who are immediately expofed to the original atmofphere, but that this contagion may certainly be communicated by the clothes of perfons
who

who have for any length of time been confined in
it; and, what is still more surprising, even when
the persons themselves have suffered no injury, nor
had any disease in consequence of such exposure.

This fact being ascertained, we cannot wonder
if those who are seized with the jail fever, owing
to such communication, should during their illness
generate a contagious vapour; but, however para-
doxical it may appear, I have never observed that
the sick propagated the disease so readily, as the
bodies and clothes of those who, though well, had
been long confined in the original atmosphere.
From my own experience also, I am led to conclude,
that there is little risk of receiving the contagion
from dead bodies, even from dissecting them, provided the surgeon does not cut himself during the
dissection, the consequence of which has generally
proved fatal.

There are several other circumstances, worthy of
notice, that increase or diminish the facility with
which contagion is communicated. Unless where
contagion is very powerful, it is seldom propagated
in the open air; I knew only one instance of this
at Winchester. It is much more certainly communicated in a room, and especially if there is a current of air, from the contagious person to others
capable of being affected. A moist atmosphere* is
also more favourable to the communication of contagion than a dry one. A contagious person becomes greatly more so, if his clothes are wet, and
his

* Moisture appears not only necessary to the production
of putrid contagion, but it would seem to be the medium
also by which it is communicated; it is well known, that
the plague ceases, in Syria and Egypt, during the prevalence of certain drying winds; and its almost entirely disap-
pearing during the winter, at Moscow, was probably owing
to the same cause; viz. the dryness of the atmosphere.

his body heated by exercife, fo as to be in a ftate of perfpiration. Thofe moft fufceptible of contagion are, young perfons, particularly if they come directly from a pure air into the infected atmofphere; perfons whofe minds are oppreffed with fear or anxiety; or who have been weakened by previous illnefs; even thofe who have been fatigued, or are fafting, more readily than others whofe ftrength has not been impaired, or which has been again recruited with food. It has been farther remarked, that perfons who have iffues are feldom affected by contagion.

Of the Effects of putrid Contagion on the human Body.

PUTRID matter, in whatever way generated, if in fufficient quantity, has always fome deleterious effect; or, in other words, acts as a poifon upon the body. It is true, that the human ftomach, and ftill more remarkably, the organs of digeftion of certain animals, have the power of counteracting the feptic tendency; but this power, in our ftomachs at leaft, is very limited; and when any matter, whether generated in the body or introduced from without, has acquired a degree of putridity beyond this, it occafions naufea, vomiting, purging, great oppreffion at the region of the ftomach, and often a fever, either of the intermittent, remittent, or more continued kind. Putrid matter, directly introduced into the fyftem by means of a wound, caufes fwelling and inflammation of the lymphatic glands, often terminating fuddenly in gangrene, along with the fymptoms of a fever, greatly refembling the hofpital or jail fever: the fame proftration of ftrength, tremors, anxiety, headach, and

delirium;

delirium; with the fame irregularity in the pulfe, and, if the difeafe continues, it induces thofe appearances of the fkin, hemorrhages, and other fymptoms, that indicate a relaxation of the folids, and refolved crafis of the blood. The fevers that arife in confequence of expofure to putrid vapour or contagion, affume a variety of types and forms, according to the various circumftances of combination, degree of putridity, feafon of the year, conftitution of the patient, &c. But they, as well as the preceding, will be found to have many fymptoms in common, and fimilar to the jail and hofpital fever : and in reality all the fevers of this clafs, from the flighteft vernal intermittent to the true plague, are in my opinion only different fhades or varieties of the fame difeafe, and productions of one common caufe, viz. putrefaction. I fhall not, however, profecute this fubject farther at prefent, as I have treated it more fully in another work, which, fhould I hereafter have leifure to complete, I hope to render not altogether unworthy of the public eye.

The contagion then of the jail or hofpital fever, may juftly be confidered as one of the moft fubtil and powerful vapours of the putrid kind ; and, confequently, its immediate and deftructive effects upon the body, are not to be wondered at. In ordinary cafes of fever, the vital principle is roufed into action, and Nature is commonly fufficient of herfelf to remove the morbid caufe; but here, as in the real peftilence, the contagion introduced into the body, feems to act as a narcotic poifon upon the heart and nervous fyftem, fuppreffing the principle of life, inftead of roufing it to the conflict. In this diftemper therefore, where nature can do fo little, and even art, unlefs immediately called to her affiftance, is equally unavailing, it is of the moft confequence for

us

us to know whether the contagion cannot be prevented or deftroyed.

Of the Means of preventing, and of deſtroying the Jail Contagion.

As we are perfectly acquainted with the caufes
of the jail contagion, we could certainly prevent its
formation, provided the means of doing ſo were always in our power; but as we cannot command
thefe, our next object is to endeavour to correct, or
deftroy it, when formed. As a knowledge of the
nature and origin of the jail contagion naturally led
to the proper and effectual means of correcting or
deſtroying it, ſo, on the other hand, the means that
have been fuccefsfully employed to deftroy it, afford
the moſt convincing evidence of its true nature.

The various means hitherto employed for deſtroying contagion, may be arranged under two diftinct
heads, or claffes, viz. the Phyfical and the Chemical.

All contagions, whether fpecific or putrid, are
either checked or completely deftroyed, by the ex-
tremes of heat and cold; and from a free expofure
to air and water, are ſo diluted or diffolved, as to
lofe their noxious quality. Heat and cold then,
with air and water, may be looked upon as phyfical agents, which, under certain circumftances, are
effectual in blunting or deftroying contagion. A degree of heat, nearly that of an oven, is found neceffary for the complete deftruction of contagion,
but as this degree of heat is incompatible with animal life,* its application is folely confined to the

D purifying

* A great heat, like that of an oven, fuch as would prove
deſtructive to all animal life, effectually deftroys this infection
in all fubftances which can be for fome time expofed to it.
Vide Lind's Obfervations on the Jail Diftemper, Ann. 1779

purifying of fuch clothes, furniture, &c. as cannot be injured by this treatment. But, although the degree of heat requifite for the complete deftruction of contagion can only be ufed for one particular purpofe, heat and fire, judicioufly managed, may, in various ways tend to leffen the power, or to check the progrefs of this pernicious vapour : for as clofe-nefs and dampnefs are favourable to the production and fpreading of contagion, drying and rarefying the air, by counteracting thefe, muft, fo far at leaft, be proper antidotes. But, independent of thofe effects of heat, an open fire, efpecially where the fuel is burnt in a narrow flue, is of great benefit ; for, by confuming a portion of the air, it caufes a more fenfible renewal of it, and, in fact, is one of the beft ventilators. In employing fire and heat, however, care muft be taken not to increafe the heat in the apartments of the fick, as this would prove more hurtful to them, than the drying or renewing of the air could be advantageous.

The degree of cold neceffary to deftroy contagion is probably, like the degree of heat, inconfiftent with life ; and, therefore, although we hear of con-tagion having been checked or fuppreffed by cold, there are few inftances, if any, of its being com-pletely deftroyed. Befides, as it is not in our power to employ cold at pleafure, the queftion refpecting its effect, of whatever importance it may be to the pathologift, is of little confequence to the practical phyfician.

That noxious vapours are hurtful only when con-centrated, and are harmlefs when diffufed, are facts or data univerfally admitted ; and it is upon this principle, that clothes, bedding, or other matters to which contagion adheres, are purified, or lofe their deleterious quality, by expofure for a fufficient length

length of time to the open air, or to a current of water ; but, as the time requisite for this mode of purification is uncertain, and as contagious clothes, goods, &c. cannot always be exposed in a proper manner,* we are commonly under the necessity of having recourse to those more expeditious means of purification which chemistry affords, and which I shall next examine.

The chemical means hitherto employed for destroying contagion, are the following :

Burning sulphur with charcoal.
——— with arsenic.
nitre.
gunpowder.
portfire.
tar.
tobacco.
wood.
Boiling vinegar.
——— with camphire.
tar.
Washing with vinegar.
White-washing.
Painting.

The vapour produced by the burning of sulphur, is known to be the volatile vitriolic or sulphureous acid, one of the most powerful of the mineral
kingdom,

* Dr. Lind has very justly remarked, that no ventilation or admission of air into prisons or hospitals, can remove or destroy contagion when once it is present. The same may be said of water. But though neither one nor the other under those circumstances can destroy contagion, both may be usefully employed in blunting its force, and in preventing the spreading of the disease.

kingdom, the effect of which in deftroying contagion has been long eftablifhed ; but as it affects,
even in fmall quantity, the refpiration of animals,
inducing fuffocation and death, it can only be employed for fumigating clothes, furniture, or empty
apar.ments. When burnt with charcoal, in the
common way, we obtain not only the fulphureous
acid, but alfo the carbonic, or fixed air, which,
though it can have little influence on contagion,
renders the common air lefs fit for refpiration ; a
circumftance hardly deferving attention where the
fulphur is burnt in a fumigating room, or a place
fet apart exprefsly for the purpofe of fumigation,
but which is of great importance when fulphur is
burnt between the decks of fhips, or in hofpital or
prifon wards, where men are foon afterwards to be
lodged. The occafional addition of arfenic feems
to have been made by Dr. Lind, with a view of
increafing the deleterious* quality of the vapour ;
but it appears unneceffary, as the fulphureous acid
is of itfelf fufficiently powerful for deftroying contagion ; befides, I doubt much, if the vapour of
arfenic is not too heavy to rife with the acid of fulphur.

Burning or deflagrating nitre.—Having had fome
experience of the efficacy of the nitrous acid in deftroying contagion, and being fenfible of the difad-
vantage of fumigating hofpital or prifon wards by
burning fulphur with charcoal, as was commonly
practifed, I refolved to employ nitre, inftead of fulphur, at Winchefter ; never doubting that I fhould
obtain,

* It was an old and very generally received opinion, that
contagious difeafes, as well as fome infections, were caufed
by infects, and therefore Dr. Lind might think, that the
moft deleterious vapour would prove the moft effectual in
deftroying contagion.

obtain, by deflagrating nitre, a portion of nitrous
acid, as well as the dephlogifticated nitrous air or
oxygene; but a farther acquaintance with chemiftry
convinced me of my miftake, and that the deflagration of nitre never produced any nitrous acid. It
is therefore evident, that deflagrating nitre in the
prifon and hofpital wards at Winchefter, could
have no effect in deftroying contagion, and no farther effect in purifying them, but fo far as it fur-
nifhed a quantity of oxygene, or air much purer
than the common air of the atmofphere.

I obferve, in Dr. Rufh's publication on the yellow fever of Philadelphia, that the phyficians of
that city lately fell into the fame miftake that I
formerly did, *viz.* deflagrating or burning nitre as
a means of deftroying contagion.

Burning gunpowder.—If there is no nitrous acid
obtained by burning or deflagrating pure nitre, we
cannot expect to procure any from burning gunpowder,* either wet or dry†. The charcoal in the
compofition poffibly yields a fmall quantity of carbonic acid, whilft the fulphur, uniting chiefly with
the alkaline bafis of the nitre, forms a hepar ful-
phuris, as the water ufed in wafhing a gun plainly
fhews.

Burning portfire.‡—This compofition of fulphur,
nitre, and charcoal, has likewife been employed§
for deftroying contagion; and as the fulphur in this

is

---

* Gunpowder confifts of feventy-five parts of pure nitre,
fifteen and a half of charcoal, and nine and a half of fulphur.

† Next to the fmoke of wood, for purifying a tainted air,
I efteem that of gunpowder. This I often ufe, as being
quite inoffenfive to the lungs, &c. Vid. Lind on Fevers
and Infections, p. 51.

‡ Portfire is made of one half fulphur, one fourth nitre,
and as much charcoal.

§ Vid. Chifholm on the Weft India Fever.

is the predominant article, it will perhaps furnish
some sulphureous acid, though I should apprehend
not a sufficient quantity to be effectual in destroying
contagion.

Burning tar.*—The use of tar seems natural
enough to sailors, who may be supposed partial to
what they are constantly accustomed; but, if we
examine the subject with attention, it is evident
that the vapour arising from tar, whether burnt or
boiled, must be a weak agent against contagion.
The empyreumatic oil can be of no service but by
opposing one disagreeable smell to another, whilst
the ligneous acid, at best a weak one compared
with the mineral acids, is in great measure destroy-
ed by burning, and is so diffused in the vapour of
boiling tar, as to prevent entirely any effect which
this acid, in a more concentrated state, might otherwise produce.

Burning tobacco.—There is an ancient prejudice
respecting the antipestilential quality of tobacco,
founded, I believe, on a tradition which is entirely
void of truth, that the plague never entered a tobacco shop. Dr. Lind however seems to have had a
high opinion of it,† but upon what this was founded I cannot pretend to say, as the smoke of tobacco, so far as I can perceive, has no advantage over
the smoke of any other vegetable matter.

Burning

---

* By smokeing this ship (Revenge) well with the vapour
of tar, the infection had abated. Vid. Lind, p. 2.

† When prisoners can be removed, the infection will most
effectually be extinguished by their removal to another prison, and, after thoroughly cleaning the infected one, to fumigate with the smoke of tobacco, &c. Vide Dr. Lind's
Health of Seamen, p. 337—Dr. Lind had so high an opinion
of the power of tobacco, that he advised the buying up all
the damaged tobacco, to be employed for this particular
purpose.

Burning wood.—The fmoke* of a wood fire was reckoned, by Dr. Lind, one of the moſt powerful means of deſtroying contagion, and he gives feveral examples where it was fuccefsfully employed. I might perhaps remark that fome of thefe examples he had from perfons who were not fuch accurate obfervers as himfelf; I fhall not however call them in queſtion, as I think the advantage fuppofed to be immediately derived from the fmoke of wood, may fairly be afcribed to other caufes. In the firſt place, the fmoke of wood confiſts principally of foot, or of inflammable matter unconfumed, with fome carbonic acid, neither of which can have any effect on contagion ; whilſt the ligneous acid is in very fmall quantity, too fmall certainly to be of much fervice. But we know, that where there is fmoke there is heat, and that where there is much fmoke, in places where people are prefent, a free admiſſion muſt be given to the air ; two circumſtances which have confiderable influence in weakening the virulence, and in preventing the fpreading of contagion.

Boiling vinegar.—Vinegar† has, at all times, been confidered as the grand antidote to contagion, though

---

* A judicious application of fire and fmoke, is the true means appropriated for the deſtruction and utter extinction of the moſt malignant fources of difeafe. Again it hath been experimentally found, that the fmoke of a wood fire ferves not only to leſſen the force or violence of fuch poifons, but is alfo an excellent protection againſt their being conveyed. Vide Lind's Papers on Fevers and Infection. Paper 1. p. 49.

† The cafcarilla bark, when burning, gives a moſt agreeable fcent to the chambers of the fick, and fo is at leaſt an excellent prefervative, and may prevent bad fmells from taking effect. The ſteam of boiling camphorated vinegar is ſtill more powerful for this purpofe. Vide Lind on Fevers and Infection. p. 51.

though I believe it to be one of the moſt trifling
means that has ever yet been employed. I have
never once obſerved the ſmalleſt benefit from its
uſe; and have known many fatal examples of contagion having been communicated where it was conſtantly employed. But although the ſteam of boiling vinegar can be of no advantage in deſtroying
contagion, yet, as the ſmell of it is grateful to the
ſick, it may for that reaſon be uſed about their perſons; and. when camphire is diſſolved in it, the
ſmell is ſtill more agreeable and reviving.

Waſhing the furniture, floors, walls, &c. with
vinegar, I conſider as little better than waſhing
them with ſimple water. The ſame may be ſaid of
white-waſhing, as the lime and ſize can have no
particular effect. Oil painting, another mode of
purifying apartments, has little advantage over the
preceding; not to mention the expenſe and inconvenience attending it.

But enough has been ſaid to ſhew the general
want of chemical knowledge, apparent in all the
methods hitherto propoſed for deſtroying contagion,
and more eſpecially, the inefficacy of the methods
employed in places and ſituations from which people could not be removed; I ſhall now proceed to
a more agreeable taſk, and explain thoſe improvements, which a more accurate chemiſtry, and a
long attention to the ſubject, have ſuggeſted to me,
and relate ſome experiments which I made, with a
view to aſcertain the efficacy of the nitrous acid,
and the ſafety with which it may be uſed, where
people are neceſſarily preſent.

The mineral acids, particularly when in a ſtate
of vapour, with the different gaſes or permanently
elaſtic fluids produced by them, are probably, excepting fire, the moſt powerful agents in nature,

and

and the fource of an infinite number of the different forms of matter obfervable in the mineral kingdom, and which are conftantly undergoing frefh
changes, from their various combinations, and de-
compofitions. But their power is not confined to
the mineral kingdom; they are known to have
great influence likewife over putrefaction, and thofe
other fpontaneous changes which vegetable and
animal matter, deprived of life, undergoes; and
therefore, if the jail contagion, as I have endeavoured to prove, is a vapour produced by putrefaction, there cannot be a doubt that the mineral
acids will prove effectual in deftroying it. So far
we may reafon *a priori*; but let us next confult experience, a lefs fallible guide. From this it appears, that the volatile vitriolic or fulphureous acid,
the only one hitherto made ufe of, proves effectual
in deftroying contagion; although owing to its deleterious quality, it cannot be employed, except in
fituations from which people can be removed.
But, are the other mineral acids in a ftate of vapour equally dangerous with the fulphureous? and,
are they equally effectual in deftroying contagion?
To the firft of thefe queftions I can give a pofitive
anfwer; to the fecond I can give one that, at leaft,
is highly probable.

In the firft place, I can fafely affirm, that the
nitrous acid may be employed in very great quantity without rifk, and even without the fmalleft inconvenience; and, that it is effectual for the deftroying of contagion, I have every reafon to believe, not only from analogy, but from experience.
I had frequently ufed the nitrous acid, as a fumigation, in hofpital wards, and in the private apartments of the fick, without perceiving any unpleafant effect from it; but, to afcertain with more

E   precifion

precision a fact of this importance, I made the following experiments; in the conducting of which, Mr. Hume of Long-acre, a very ingenious man, and an excellent chemist, was so obliging as to favour me with his assistance.

We put a mouse, confined in a wire trap, under a glass cylindrical jar, capable of holding about 25 pints beer measure, or 881 cubic inches; the jar was inverted upon wet sand, contained in a flat earthen trough or pan; it was then filled with the fumes of the smoking nitrous acid, introduced by means of a crooked glass tube, until the animal could not be very distinctly perceived. The mouse was kept in this situation for a quarter of an hour, when the jar was removed, and the animal exposed to the open air; it immediately ran about the wire trap, as usual, and had not the appearance of having suffered the slightest inconvenience from its confinement. After a few minutes, the mouse was again put under the glass jar, which was now filled with the vapour of pure nitrous acid, detached from nitre by the vitriolic acid. It remained much about the same time as before, and when the jar was removed, seemed perfectly well.

We repeated the same experiments with a green-finch, only with some little variation in the manner. We placed, on a table covered with green baize, a brown earthen vessel or pan, containing heated sand; in this was put a glass saucer, with about half an ounce of strong vitriolic acid; above which we placed the bird-cage, supported with some small pieces of wood laid across the pan; then, adding a drachm or two of nitre, in powder, to the vitriolic acid, we covered the whole with the glass jar. The nitrous acid rose in such quantity, that, in a very little time, the bird seemed as if in a

cloud

cloud or fog. We kept it in this fituation fifteen minutes, by which time the cloud had difappeared, and the acid was in part condenfed on the fide of the glafs jar; during the whole time the bird neither panted, nor appeared to fuffer any uneafinefs, from the atmofphere in which it was confined. We made trial alfo of the marine acid, by adding common falt, inftead of nitre, to heated vitriolic acid: during this experiment, the bird appeared to be, now and then, fomewhat uneafy, and opened its bill; but at the end of fifteen minutes, upon removing the jar, it hopped about as lively as before. We then expofed the bird to the fumes of fulphur, burnt with an eighth part of nitre; it immediately gave figns of uneafinefs, opened its bill, and feemed to pant for breath in fuch a manner, that we were afraid to cover it with the glafs jar. We likewife made trial, in the open air, of the oxygenated marine acid;* for, as this is fo extremely deleterious, we did not think it fafe to expofe ourfelves to the vapour of it in a room, nor did we venture to expofe the bird to it in any other way but in the open air, and even there it appeared to fuffer very much.

Having made trial of the effect of the different mineral acids, in a ftate of vapour, upon animals, we determined to render the experiment ftill more conclufive, by trying what effect they would have on ourfelves. With this intention, we filled the
room

---

* The oxygenated marine acid is a difcovery of the famous Scheele, and has been recommended by Berthollet and Chaptal, two French chemifts, for the purpofe of bleaching. I am informed that it has alfo been lately ufed in France to deftroy contagion, but the particular circumftances, and manner of its application, I have not yet learnt.

room* in which we were with the fumes of nitrous acid, (obtained by mixing nitre with heated vitriolic acid, in the manner already defcribed) until the different objects became fomewhat obfcure, by a kind of fog or mift produced. The fire irons and fteel fender, loft their polifh, and the vapour arifing from a bottle of aqua ammoniæ puræ, placed at fome diftance from the table, was evidently neutralized, as it iffued from the bottle, by the vapour of the nitrous acid.

Mr. Hume and I remained in the room the whole time, without perceiving the flighteft inconvenience; the fumes did not excite coughing, nor affect the eyes, in the way the fmoke of wood commonly does, even when I held my head over the glafs faucer, and breathed them immediately arifing from it. We made trial likewife of the effect of the marine acid, which we found more pungent and ftimulating than the nitrous; but, though it excited coughing, it did not caufe that conftriction of the windpipe, and tightnefs at the cheft, with the fenfe of fuffocation, which is immediately induced by the volatile vitriolic or fulphureous acid. Indeed we were imprudent enough to try how far we could breathe this laft, but I was inftantly obliged to run to the window for air, from the fenfe of conftriction, and of fuffocation, which it occafioned. We likewife tried the effect of the mixed fumes of the marine and nitrous acid, a kind of volatile aqua regia, which we found more pungent than the marine acid by itfelf. As for the oxygenated marine acid, perceiving the effect of it on the bird,

---

* The room in which we made the experiments was a fmall parlour 13 feet by 10, and 8 feet high; or about 1040 cubic feet.

bird, and knowing how extremely dangerous it is, we did not venture to go very near it.

From the preceding experiments, the different acid vapours, in refpect to the fafety with which they may be breathed, may be arranged in the following order :

1ft. The vapour of nitrous acid, arifing from nitre decompofed by vitriolic acid.

2. Ditto—of nitrous acid in its fuming ftate, or when the nitric acid is mixed with nitrous gas.

3. Ditto—of marine acid, arifing from common falt, decompofed by vitriolic acid.

4. Ditto—of nitrous and marine acids, obtained from the decompofition of nitre and common falt by vitriolic acid.

5. Ditto—of fulphur, burnt with an eighth part of nitre.

6. Ditto—of fulphur, burnt with charcoal.

7. Ditto—of oxygenated marine acid,* obtained by putting manganefe to marine acid.

As the firft vapour is perfectly harmlefs, in any quantity in which it may be required, it is evidently the moft proper to be employed in all fituations where people are neceffarily prefent ; and if it fhould prove efficacious in deftroying contagion, of which I have not the fmalleft doubt, it is the *defideratum*,† fo much fought after by Dr. Lind ; but which

* The oxygenated marine acid is obtained, by diftilling marine acid from manganefe, but may alfo be procured in fmall quantity, by putting manganefe to heated marine acid, or by gradually adding a mixture of manganefe and fea-falt to heated vitriolic acid.

† A certain method therefore of deftroying infection in places from whence perfons cannot be removed, is a *defideratum* not yet obtained in phyfic. I have propofed and tried

which he confesses, with his usual candour, he never could find out.

The second, though more pungent than the first, may I believe be employed with the greatest safety; at least, I have never observed any inconvenience from using it. But, as it cannot so easily be procured in considerable quantity, and is attended with greater inconvenience and expense, I have of late years only made use of the first.

Our experiments likewise warrant us to affirm, that the third, or marine acid, though more stimulating, and more apt to excite coughing, than the nitrous, may be safely used, at least in a moderate quantity, where people are present; and where nitre cannot be had, I should have no hesitation in employing it.

Of the fourth I can say but little, only that, in breathing it, I perceived it more pungent than the pure marine acid; and therefore, unless it should be found to possess superior efficacy in destroying contagion, I would not employ it where there are people present.

As the fifth never can be used with safety where there are people present, its use must be solely confined to fumigating empty apartments, clothes, furniture, &c.

The sixth should never be employed, as the carbonic acid may do harm, and never can have any effect on contagion.

Of the seventh I have no particular knowledge, only that it is extremely deleterious, and I believe extremely powerful; but whether it has more effect

on

many things for this purpose without success. Vide Lind's Observations on the Jail Distemper. Edit. published in Oct. 1779.

on contagion than the other mineral acids, experience only can determine.

Having now fully proved that the nitrous, and possibly also the marine acid, obtained in the manner already described, may be employed with perfect safety in hospital and prison wards, whilst the people remain in them, I shall, in the next place, relate how far my experience goes to ascertain the efficacy of those acids in destroying contagion.

From all the information I can procure, I do not find that any person has ever made use of the nitrous acid to destroy contagion but myself; for, as this acid is not produced by the deflagration of nitre, or of gunpowder, the employment of these cannot be considered as an instance to the contrary. I formerly mentioned, that I had employed the nitrous acid in two different forms; either the vapour arising from the yellow or smoking nitrous acid, which is a mixture of the acid with nitrous gas, or the more pure nitrous acid, detached from nitre, decomposed by the vitriolic acid. In one or other of those forms I have used it, both in hospitals and in private practice, for sixteen or seventeen years past; and have had the satisfaction to obtain the most decisive evidence of its happy effect, in preventing the spreading, or farther communication of contagion.

The most highly contagious fevers that occur in our hospitals, do not affect the patients in general lodged in the same ward, but only the nurses, or those patients who assist them, or those who lie in the beds contiguous to the sick; to such persons I have frequently seen the fever communicated, and have also repeatedly prevented the farther spreading of the disease, by placing gallipots, with the fuming nitrous acid, between the beds of the sick and

of

of thofe who were not yet affected by the contagion. And, in private practice, I can declare with truth, that where the nitrous acid has been conftantly ufed as a fumigation, I have not known an inftance of a contagious fever having been communicated, even to a nurfe or an attendant.

Thefe facts will, undoubtedly, be allowed to be very ftrong evidence, with refpect to the power of the nitrous acid to deftroy contagion; ftill, however, they are liable to fome uncertainty, and I will freely confefs, that the effect of the nitrous acid, for this purpofe, cannot be faid to be fully proved, until it has been tried in fumigating tainted clothes, &c. and until its power has been found fufficient to deftroy contagion on board of fhips, and in prifons and hofpitals, where it exifts in a much higher degree than I have had occafion to fee it, excepting at Winchefter.

It will probably be afked, why I did not make a complete trial of it there? To this I anfwer, that with refpect to fumigating infected clothes, bedding, &c. I did not think myfelf warranted, efpecially on an occafion of fo much importance, to make trial of an uncertain remedy, when a certain one was in my power. As to fumigating the prifon and hofpital wards, it was evidently my intention to have employed the nitrous acid, but I was miftaken in the means I took to procure it, and have not fince had a proper opportunity of repeating the experiment.

The effect of the marine acid, in a ftate of vapour, on contagion, I have not yet had occafion to try, but have no doubt that it will be found of fufficient efficacy for deftroying it; and, from the foregoing experiments, it is evident that, though not fo mild or fafe as the nitrous acid, it may be

ufed,

uſed, in a moderate quantity, even where people are preſent. The only purpoſe to which I have applied it, has been, when properly diluted, to waſh the hammock poſts, bedſteads, and furniture ; alſo the floors, and walls, of the apartments of the ſick :* and I am perſuaded that, even in this way, it was extremely ſerviceable, certainly more powerful than the moſt concentrated vinegar.

I ſhall now conclude this ſubject with a few practical rules or obſervations, which may be looked upon as corollaries, or inductions, from the preceding experiments.

The well known efficacy of the ſulphureous acid, in deſtroying contagion, is a ſufficient reaſon for our continuing to uſe it as a fumigation for clothes, furniture, &c.

The nitrous acid, being attended with no riſk or inconvenience to the reſpiration, and appearing, from our experience, of ſufficient efficacy to prevent the farther ſpreading of contagion, ſeems the proper antidote to be applied, in all ſituations where perſons are neceſſarily preſent, and is, in ſhort, the *deſideratum* ſought after by the benevolent Dr. Lind.

For purifying empty hoſpital or priſon wards, and ſhips, I ſhould alſo prefer the nitrous acid to the ſulphureous ; as I believe it to be equally efficacious ; its vapour is more volatile and penetrating ; and it does not leave the diſagreeable ſmell which ſulphur does. But, for this particular object, I think it would be adviſeable to make trial

F   alſo

---

* The waſhing the hammock poſts, walls, and floors of the priſon wards with the diluted marine acid, and the removal of all clothes, bedding, &c. proved completely effectual for deſtroying the contagion at Wincheſter ; as it is now apparent, that the burning or deflagrating of nitre could contribute nothing to the ſucceſs.

alfo of the marine acid, and of the mixture of nitrous and marine acids, as I am convinced of the
efficacy of all the mineral acids for deftroying contagion, and our experience is not yet fufficient to determine their relative advantages, and difadvantages.

To obtain the nitrous, or marine acid, in a ftate
of vapour, the method is extremely fimple. It confifts in decompofing nitre, or common falt, by means
of heated vitriolic acid, which may be done as
follows :

Put half an ounce* of vitriolic acid into a cruci-
ble, or into a glafs or china cup, or deep faucer;
warm this over a lamp, or in heated fand, adding
to it from time to time fome nitre or common falt :
thefe veffels fhould be placed at twenty or thirty
feet diftance from each other, according to the
height of the cieling, or virulence of the contagion.
In hofpitals, or prifons, the lamps, or veffels containing heated fand, may be placed on the floor;
on board of fhips, it will be better to hang them
to the cieling by waxed filk cords. The fumigating
lamps, which I have feen at **Moyfer's**, in Greek-
ftreet, Soho, a great number of which I was told
have been fold to the navy, may be employed for
this purpofe; although they would anfwer much
better, if the faucer was deeper, and if, inftead of
a place for a lamp, there was a box proper for containing hot fand, in which the faucer might be
placed.

As fumigating with nitrous acid is attended with
no inconvenience, and as the procefs is fo fimple,
and

---

* As the quantity of vapour depends, in fome meafure,
on the furface, I think it better to have the vitriolic acid put
in a number of fmall veffels, than in one or two large ones ;
befides, in this way, it has the advantage of being diffufed
more readily in any given fpace.

and the materials fo cheap, it fhould, as a means of prevention, be employed for fome hours every day in tranfports having troops on board, and in crouded hofpitals; and, if there is any appearance of contagion, the fumigation fhould be executed with more care and attention, and the vapour confined for feveral hours at a time. Fumigating veffels, or lamps, fhould alfo be placed contiguous to the hammocks, or beds, of perfons affected with any contagious or putrid diftemper, whether fever or dyfentery.

By taking fuch precautions, a great deal of mifchief would probably be prevented, and a ftop put, in the beginning, to one of the moft fatal calamities* that ever afflicted mankind.

---

* The late dreadful mortality in the Weft-India iflands, occafioned by a contagious fever imported from Boulam, has made too deep an impreffion on the minds of the people of this country to be foon forgotten, and every exertion on the part of the executive government will no doubt be made to prevent a repetition of the fame tragedy.

SINCE

SINCE writing the above, I have had the plea-
sure of feeing the laſt publication, and, as I imagine, the lateſt improvements, of the French chemiſts and phyſicians on the ſubjeſt of contagion,
and on the proper means of deſtroying it. It is in-
titled, " *Inſtruſtion, ſur les moyens d'entretenir la*
" *ſalubrite, et de purifier l'Air des Salles, dans les*
" *Hopitaux militaires de la Republique, fait au Con-*
" *ſeil de Santé le 5 Ventoſe, l'An 2d de la Repub-*
" *lique Françaiſe une et indiviſible.*"

This *inſtruſtion*, or *memoire*, is divided into three
parts. The firſt relates ſolely to the means of
cleanlineſs; the ſecond to what are called the me-
chanical means; and the third to the chemical.
The two firſt parts contain nothing new or intereſting; the third is of the greateſt importance to medical ſcience, and particularly ſo to me, as it fur-
niſhes a proof of the accuracy of ſome of the preceding experiments, and is a complete confirmation
of the opinions I have long entertained reſpeſting
the nature of contagion, and the power of the
mineral acids to deſtroy it.

The French phyſicians, inſtruſted by that excellent chemiſt *Le Citoyen* Guiton, better known by
the name of Monſ. de Morveau, of Dijon, have
lately made trial of the marine acid in their hoſpitals, and have found it equally effeſtual in deſtroying contagion as the ſulphureous, and, as being
more volatile, perhaps even preferable for the purpoſe of purifying hoſpital wards. They alſo remarked that, in a ſmaller proportion, it may be
ſafely uſed in hoſpital wards, even when people are
preſent.* The French phyſicians however have
not

---

* My experiments ſhewed the ſame thing.

not employed the nitrous acid, nor made any trials of its effect on contagion ; neither do they appear to have fufpected that the power of deftroying contagion was a quality inherent in all mineral acids ; and probably, to a certain degree, in all acids, under certain circumftances. Although their experience of the effect of the marine acid, together with my obfervations on that of the nitrous, feems to eftablifh the fact beyond the cavil of fcepticifm itfelf.

Their method of obtaining the marine acid is the fame that I took to procure the nitrous ; they either employed the fuming marine acid, or the acid detached from its alkaline bafis by vitriolic acid, ufing a confiderable degree of heat for that purpofe.* They likewife, upon the fuggeftion of M. Fourcroy, recommend adding a fmall quantity of the oxygenated marine acid ; but, as they do not pretend to fay that they have had any experience of the fuperior efficacy of this, and as the common marine acid has been found to anfwer the purpofe, I do not fee any reafon for making fo hazardous an addition.

Another chemical procefs for purifying foul air in hofpitals, recommended in this *inftruction*, deferves our notice. It confifts in placing, at different diftances in the hofpital wards, veffels with lime water, for the purpofe of abforbing carbonic acid or fixed air. I am inclined, however, to believe, that this advice is more the refult of chemical theory than of practical obfervation ; for I do not fuppofe that carbonic acid is ever prefent, (where there is a free admittance of air,) in fufficient

* The reader will find at the end an account of their procefs.

cient quantity to prove hurtful; at leaft, it can only affect the breathing, and has nothing in common with contagious vapour.

The French phyficians appear to me to have fallen into a confiderable miftake on this fubject, in taking the quantity of carbonic acid prefent,* in an hofpital, as a teft of the quantity or malignity of contagion, when, in reality, they are two things totally diftinct from each other. The firft, or carbonic acid, is a conftituent part of the common or atmofpheric air, which is greatly increafed by the refpiration of animals, and by burning candles, lamps, &c. and, when in too great quantity, extinguifhes flame, and animal life : the other has no relation with the compofition of the atmofphere, never affects refpiration, but is produced by putridity, and excites fever.

* The method propofed by the French phyficians, for afcertaining the quantity of carbonic acid prefent, is fimple and ingenious. Take two phials; let one be filled with common water, the other with lime water. At the place where you want to try the purity of the air, empty the phial of common water, then, filling it half full with lime-water, and corking it, fhake the phial for fome time : the quantity of fediment fhews the proportion of carbonic acid. But, to render the preceding experiment conclufive, the height from the ground at which the air is taken fhould be ftated, otherwife we are liable to great fallacy.

Extract

Extract from the " *Instruction, sur les Moyens*
" *d'entretenir la salubrité, et de purifier l'Air des*
" *Salles dans les Hopitaux Militaires de la Répub-*
" *lique, &c. &c. &c.*"

" *Au nombres des moyens que la chimie a employés*
" *avec un succès que tient du prodige pour operer*
" *cette depuration, nous citerons le procedé que*
" *Guiton, (Monf. de Morveau) reprefentant du*
" *peuple, a mis en ufage en* 1773, *dans la ci-devant*
" *cathedrale de Dijon, infectée par des exhumations,*
" *au point qu'on fut obligé de l'abandonner.*

" *Ce moyen confiste à repandre dans l'atmofphere,*
" *de l'acide muriatique (acid marin) en etat de gaz*
" *degagé par l'intermède de l'acide fulphuric ; (huile*
" *de vitriol) voici le procédé pour définfecter une falle*
" *de* 40 *a* 50 *lits.*

" *Après avoir évacué les malades fur une des falles*
" *de rechange, difpofez dans le milieu de la falle*
" *vuide, dont les fenétres & les portes feront fermées,*
" *un fourneau garni d'une petite chaudière ou capfule*
" *de fer, à demi remplie de cendre tamifée fur la-*
" *quelle on pofera une capfule de verre de grès, ou*
" *de fayance méme, chargée de neuf onces de muriate*
" *de foude, (fel marin,) legérement humecté avec*
" *une demi-once au plus d'eau commune. Le feu*
" *étant allumé à la capfule echauffie, on verfera fur*
" *le fel marin quatre onces d'acide fulfurique, huile*
" *de vitriol de commerce. En un instant l'acide ful-*
" *furique agira fur le fel marin, dont l'acide fe*
" *mettra en expanfion ; l'operateur, qui fera le phar-*
" *macien en chef, ou un de fes aides, verfé dans le*
" *manuel des operations chimiques, fe retirera, en*
" *fermant la porte fur lui, et emportant la clef ; douze*
" *heures après on entrera dans la falle, on ouvrira*
" *portes*

" portes et fenêtres, pour établir des courans d'air,
" et évacuer celui qui pourroit être encore chargé
" d'acide. On donnera une plus grande latitude
" d'utilité à ce procédé en l'appliquant aux salles même
" remplies de malades, toutes les fois que les officiers
" de santé le jugeront necessaire. Ainsi lorsqu'on
" aura reconnu que l'air d'une salle est surchargé de
" miasmes animaux, et a besoin de cet excellent puri-
" ficateur, il suffira de faire le tiers du melange ci
" dessus, et même moins, et de la parcourir plus ou
" moins lentement, et dans tout les points, le rechaud
" à la main, au moment où le gaz se met en expansion.
" Lorsque la salle sera jugée suffisamment rempli
" de gaz acide muriatique, on transportera l'appareil
" dans les latrines, afin que les dernieres portions
" gazeuses que le mélange pourra continuer de fournir
" servent à neutralizer les gaz ammoniacaux putrides,
" qui se developpent continuellement dans les privés.

EXPE-

EXPERIMENT

MADE WITH THE

NITROUS FUMIGATION

ON BOARD THE

UNION HOSPITAL SHIP,

&c. &c. &c.

G

THE RIGHT HONOURABLE

EARL SPENCER,

&c. &c. &c.

My Lord,

THE general opinion entertained of your Lordſhip, in the high department, at the head of which
you are placed, is the only apology I can offer for
having taken the liberty to trouble you on the ſubject of my late publication. The immediate attention paid to this by your Lordſhip, and by the reſt
of the Lords Commiſſioners of the Admiralty, is
extremel█ flattering to me, as an individual, and
claims my warmeſt gratitude; but it is of much more
importance, my Lord, as holding out to the nation,
a well grounded confidence, that no object which
may be conducive to the public ſervice, or to the
preſervation of thoſe brave men, the pride and pro-
tectors of their country, can long eſcape your Lordſhip's notice. I have now the honor to lay before
you, and the reſt of the Lord's Commiſſioners of
the Admiralty, an account of the Experiment made

on

on board the Union, at your Lordſhip's deſire, and likewiſe of thoſe trials that were made at the deſire of the Ruſſian Admiral, and with your Lordſhip's approbation, on board ſome ſhips of his ſquadron. I conſider myſelf, in executing this taſk, as only performing a duty I owe to your Lordſhip, and which I do with the greater pleaſure, as it may poſſibly be the means of making public a diſcovery which ſhould be univerſally known; and as the only way in my power to bring forward the merit of thoſe Gentlemen, to whoſe aſſiſtance I have been particularly indebted for the fortunate iſſue of this experiment, and from whoſe reports I am enabled to preſent your Lordſhip with an account of the manner in which it was conducted, and of the particular effects it produced.

Mr. Menzies, late Surgeon to his Majeſty's ſloop the Diſcovery, was the perſon who, at my requeſt, very obligingly undertook the management of the experiment on board the Union, and it is but doing him juſtice to ſay, that I could not have found a gentleman better qualified, in every reſpect, for executing ſo important a truſt. I ſhall therefore, my Lord, without farther preface, lay before you, and the reſt of their Lordſhips, Mr. Menzies's journal, as affording a better deſcription of the experiment, ſo long as he continued to con● it, than any I can offer.

REPORT

REPORT

OF THE

EXPERIMENT

FOR

Stopping the Progress of Contagion, *as executed on board the* Union Hospital Ship, *at* Sheerness, *by Mr. Arch. Menzies.*

DOCTOR James Carmichael Smyth having been requested, by the Lords Commissioners of the Admiralty, to send a person on board the Union Hospital Ship, laying at Sheerness, to make trial of the effect of a fumigation of the Nitrous Acid, and of other means recommended by him in a late publication, for destroying Contagion, I readily engaged, upon application being made to me by some of our common friends, in the execution of an experiment which I foresaw might eventually be of much benefit to society, and particularly to that service, to which I have the honor to belong.

After

After having, therefore, received inftructions, and obtained every neceffary information on the manner of conducting the fumigation, I left London on the 24th of November, 1795, and arrived at Sheernefs the fame evening.

Next morning I waited on Admiral Buckner, the commanding officer of the port, whofe politenefs and zeal to promote the object of my journey, were equally confpicuous, and deferves my moft grateful acknowledgement.

I afterwards went on board the Union, where I produced the orders of the Admiralty to Lieutenant Quarme, the commanding officer, and Mr. Baffan, furgeon of the fhip, who received me with cordiality, and readily offered every affiftance in their power to carry on the experiment, upon the event of which not only the fafety of the fhip's company, but perhaps, their own, in great meafure, depended.

On examining the ftate of the hofpital, I plainly forefaw that frefh contagion would be daily pouring into it from the Ruffian veffels, under which difadvantageous circumftance, it would be difficult to decide on the fuccefs of our endeavours. The lower and middle gun-decks were divided into large apartments, or wards, by crofs partitions, with a free communication between each : they were extremely crouded, and the fick of every defcription lay in cradles, promifcuoufly arranged, to the number of nearly two hundred ; of which about one hundred and fifty were in different ftages of a malignant fever, extremely contagious, as appeared evident from its rapid progrefs, and fatal effects, amongft the attendants on the fick, and the fhip's company. For, from the beginning of September laft, when the Ruffian fick were firft admitted into the hofpital ; eight nurfes and two wafher-women had been at-

tacked

tacked with this fever, and of thefe three had died. About twenty-four of the fhip's company had likewife been ill of the fame diforder, and of thefe a furgeon's mate and two marines died. Upon the whole, however, the mortality had not been fo great as there were reafons to dread, from the virulence of the contagion, and malignity of the difeafe; which can only be afcribed to the great care and attention of Mr. Baffan, furgeon to the hofpital, whofe conduct in fo critical a fituation does him the higheft honor, and reflects luftre on his profeffional abilities, in the faithful difcharge of fo unpleafant a duty.

After I returned on fhore from the Union, I employed the reft of the day in collecting and fending on board fuch utenfils and materials as were required for fumigating the fhip; thefe confifted of a quantity of fine fand, about two dozen quart earthen pipkins, and as many fmall common tea-cups, together with fome long flips of glafs to be ufed as fpatulas; the other materials I had brought with me from town, viz. the concentrated vitriolic acid, and a quantity of pure nitre in powder.

On the forenoon of the twenty-fixth, I went again on board the Union. I firft ordered all the ports and fcuttles to be clofe fhut up; the fand, which had been previoufly heated in iron pots, was then fcooped out into the pipkins by means of an iron ladle, and in this heated fand, in each pipkin, a fmall tea-cup was immerfed, containing about half an ounce of concentrated vitriolic acid, to which, after it had acquired a proper degree of heat, an equal quantity of pure nitre in powder was gradually added, and the mixture ftirred with a glafs fpatula, until the vapour arofe from it in confiderable quantity. The pipkins were then carried through the wards, by the nurfes and convalefcents, who kept walking about
with

with them in their hands, occafionally putting them under the cradles of the fick, and in every corner where any foul air was fufpected to lodge. In this manner we continued fumigating, until the whole fpace between decks was, fore and aft, filled with the vapour, which appeared like a thick haze.

I however proceeded in this firft trial flowly and cautioufly, following with my eyes the pipkins in every direction, to watch the effect of the vapour on the fick, and obferved that at firft it excited a good deal of coughing, but which gradually ceafed, in proportion as it became more generally diffufed through the wards; this effect appeared indeed to be chiefly occafioned by the ignorance or inattention of thofe who carried the pipkins, in putting them fometimes too near to the faces of the fick, by which means they fuddenly inhaled the ftrong vapour, as it immediately iffued from the cups.

In compliance with Doctor Smyth's requeft, the body-clothes and bed-clothes of the fick were, as much as poffible, expofed to the nitrous vapour during the fumigation; and all the dirty linen removed from them was immediately immerfed in a tub of cold water, afterwards carried on deck, rinfed out, and hung up till nearly dry, and then fumigated before it was taken to the wafh-houfe: a precaution extremely neceffary in every infectious diforder. Due attention was alfo paid to cleanlinefs and ventilation.

As the people were at firft very awkward and flow, it took us about three hours to fumigate the fhip; in about an hour after, the vapour having entirely fubfided, the ports and fcuttles were thrown open, for the admiffion of frefh air. I then walked through the wards, and plainly perceived that the air

air of the hospital was greatly sweetened, even by this first fumigation.

Next morning the ship was again fumigated, beginning with the lower deck, and the people employed being now better acquainted with the operation, were more expert, and finished the whole in about an hour's time; in an hour afterwards, the vapour having entirely subsided, the fresh air was freely admitted into the hospital.

This day the sand was made hotter, and the fumigation was of course much stronger, yet the patients suffered no other inconvenience from it than a little coughing, and even that was not near so general as the day before.

Twelve pipkins were found sufficient for fumigating the lower deck, ten for the middle gun deck, two for the ship's company's bed-room, two for the marines' bed-room, and one for the washing place; in all twenty-seven pipkins. Consequently, about fourteen ounces of the vitriolic acid, and as much nitre, were expended in the forenoon; but, in the evening, as every place was so close, and the fresh air could not be afterwards so freely admitted, it was not thought necessary to employ so many pipkins; so that little more than half the quantity of the fumigating materials used in the morning, was generally found sufficient for the evening's fumigation.

The pleasing and immediate effect of the fumigation, in destroying the offensive and disagreeable smell arising from so many sick crouded together, was now very perceptible, even to the nurses and attendants. The consequence of which was, that they now began to place some degree of confidence in its efficacy, and approached the cradles of the infected with less dread of being attacked with the disorder; so

H   that

that the fick were better attended, and the duty of
the hofpital was more regularly and more cheerfully
performed. In fhort, a pleafing gleam of hope feemed now to caft its cheering influence, over that general defpondency which was before evidently pic-
tured in every countenance, from the dread and hor-
ror each individual naturally entertained of being,
perhaps, the next victim to the malignant powers of
a virulent contagion.

On the twenty-eighth, the fumigation was repeated morning and evening, in the fame manner as on
the preceding day, and with the fame pleafing effect,
deftroying the offenfive fmell, and purifying the general air of the hofpital. But there was, in particular places, a conftant fource of bad fmell, which
was not eafily overcome, and which was occafioned
by the *neceffaries*. Thefe were badly conftructed,
being placed within the fhip, to the number of feven
on the lower deck, and two on the middle deck,
with fmall funnels that pierced the fides of the fhip
in a flanting direction, and generally retained the
foil, unlefs where a perfon conftantly attended to
wafh it away, a very troublefome and dangerous
office, which chiefly fell to the lot of the nurfes, and
doubtlefs tended to fpread the contagion amongft
them.

I mentioned this nuifance to the commanding officer, who told me that he viewed them in the fame
light, and that fome alterations were making, which
he hoped would remedy the evil. I therefore waited a few days the event of thefe alterations, before
I fhould make any public report on the fubject.

For the following eight days I continued the fumigation on board the Union, regularly morning and
evening, as already defcribed, without obferving any
particular occurrence different from what is already
related,

related, only that during this time, a confiderable
number of patients having been difcharged from the
hofpital, all the fpare cradles were ordered on deck,
to be fcrubbed and wafhed with the diluted marine
acid, according to the particular directions of Dr.
Smyth.

On the feventh of December, I refigned to Mr.
Baffan the further profecution of the experiment on
board the Union hofpital fhip, but before I leave
her, I muft fay, that it has already produced the
moft evident and beneficial effects, as not one of the
attendants on the fick, nor any of the fhip's company
have been attacked with the diforder fince I began
the fumigation, with the exception of one nurfe,
who fuffered a flight relapfe from fome imprudence;
an accident which Mr. Baffan informs me was very
frequent in the beginning. And as none of the fick,
who have been brought to the hofpital fince my arrival, have died, it would feem that the fumigation
has not only leffened the danger of infection, but
alfo the malignity of the difeafe.

The procefs of fumigating as already defcribed,
with the *nitrous acid*, is fimple and eafy, and although the vapour is extremely powerful and penetrating, the fick of every defcription were obferved
to bear it, with little or no apparent inconvenience,
and to a much higher degree than I could have
expected; and as it is found to purify the air from
the difagreeable effluvia, produced by fo many people crouded together in a confined fituation, it
will be peculiarly advantageous on board of fickly
fhips, where the crew, their clothes, and the fhip,
may be fumigated at the fame time without any
rifk from fire.

December

December 16, 1795.

On the fixteenth of December, I again vifited the Union hofpital fhip, and found that the fumigation had been hitherto carried on regularly twice a day, and with the fame evident advantages, in purifying the air of the hofpital, and leffening the malignity of the diforder, fo that every nurfe and attendant on the fick, went now cheerfully and confidently about their duty; without the leaft dread or apprehenfion of the contagion, by which means the fick were better taken care of and the general ftate of the hofpital was in a much more profperous way. It was therefore, from this time, deemed fufficient to fumigate only once a day.

December 23, 1795.

On vifiting the Union again on the twenty-third of December, I found the carpenters employed, from the dock yard, in making the alterations which I formerly propofed in a letter to Dr. Carmichael Smyth, refpecting the *neceffaries*, and which I was happy to find, the Lords of the Admiralty had ordered to be done upon his application.

My propofal was to remove all the *neceffaries* from the infide, and have them rebuilt on the outfide of the fhip, and by cutting down the lower edge of the fame number of port-holes, to form entrances into them from the hofpital, by which they would be equally eafy of accefs to the fick, and the nuifance would be totally removed. This I was happy to find the carpenters were now executing, and I am confident it will be attended with beneficial effects, by rendering the hofpital much fweeter, and confequently more agreeable and healthy, both to the fick and attendants.

ARCHIBALD MENZIES.

Mr.

Mr. Menzies, as is already mentioned in his Journal, having, on the 7th of December, refigned to Mr. Baffan, furgeon of the Union, the management of the experiment, I muft refer your Lordfhip, for the further detail of this bufinefs, to extracts taken from his letters, fome of which you have already feen, and which are now arranged according to the order of time in which they were written.

Mr. Baffan's conduct, my Lord, through the whole of this bufinefs, does him the higheft honour, and cannot fail to recommend him to your Lordfhip's notice. When the contagion at firft began to fpread among the fhip's company of the Union, he was importuned, by the warrant officers and others, to fend them on fhore to fick quarters, which he peremptorily refufed, faying, with the true fpirit of a Britifh failor : " It is better we " fhould all perifh, than have fuch a contagious " fever as this diffeminated in our fleet." He accordingly made application to the Commander in Chief, and not a man was fent out of the fhip. His humanity and care of the fick, Mr. Menzies mentions in the warmeft terms of praife, and his fuccefsful treatment of them, is the beft teftimony of his profeffional abilities. His zeal and attention, in conducting the experiment, I fhall always recollect with gratitude. He and Mr. Menzies were both of them ftrangers to me until this occafion brought us acquainted ; but I muft fay, that in the whole circle of my acquaintance, I could not have found two more liberal or candid men.

Extracts

Extracts of Mr. Baffan's Letters to Dr. Carmichael Smyth.

Sheerness, December 4.

I beg leave to inform you, that we have continued to fumigate, in the manner directed, daily;
and as only one Ruffian has died fince we began, I
confider that circumftance as an early profpect of
our future fuccefs.

——————December 7. *

The fumigation is not attended with the fmalleft
inconvenience to any one, the majority of patients
being in bed when it is done, and all of them in the
wards; the cabins of the nurfes, privies, &c. are
fumigated, as well as the apartments of the marines,
and fhip's company. For two months prior to the
experiment, very few days elapfed without fome of
the attendants, or fhip's company being feized with
the fever; but fince the 26th ultimo, the day on
which Mr. Menzies began the fumigation, not one
has been attacked with the difeafe; one nurfe only
having relapfed, a circumftance very common, and
occafioned by her not taking care of herfelf. I beg
leave to inform you, that this day I began to take
charge of the bufinefs, in the abfence of Mr. Menzies, who is on board the Pamet Euftaphia to try
the experiment, (fhe having been the moft fickly
fhip) where I am certain he will take fuch meafures,
as will do himfelf credit, and you honour.

I intend

* This letter, which by fome accident was miflaid, and
confequently not inferted in the former edition, I have publifhed in the prefent, as it renders his correfpondence complete, and fhews the unremitting attention of that worthy man,
(whofe fervices the public have now unhappily loft,) to every
part of his duty.

I intend in a few days, fending you a journal from the 1ſt to the 26th of November, the day Mr. Menzies began in the Union, and another from the 26th ult. to the 11th inſt. containing the receipts, diſcharges, and deaths, by which you will be enabled to make a fair compariſon, much in favour of the means uſed, I am ſure. The dejection and melancholy occaſioned by the dread of the diſeaſe, prior to the commencement of the experiment, was evident in every countenance, and really affecting, and diſtreſſing; but the circumſtance of its being ſtopped at preſent, has diffuſed joy and cheerfulneſs, and all look forward with the hopes and expectation of ſoon becoming a wholeſome ſhip. The ſymptoms of the fever are certainly much leſs violent, and at preſent, I have very few people in a dangerous ſtate.

—— *December* 9.

We continue to fumigate the ſhip as formerly; your other inſtructions ſhall be punctually adhered to.

—— *December* 11.

I yeſterday ſent you two liſts or journals, one of the Ruſſian ſick, and the other of perſons belonging to the Union, who have been attacked with the fever; from the laſt you will perceive that very few days elapſed from the firſt importation of the diſeaſe, to the 26th ult. without ſome one or other of the attendants, or ſhip's company, being ſeized with it; but ſince that period not one has been taken ill. I intend, very ſhortly, to ſend you a brief account of the diſeaſe, the ſymptoms of which are at preſent much meliorated. I believe that the fumigation has been of great ſervice to the ſick. We have very few patients at preſent who are not in a
convaleſcent

convalefcent ftate, and there is every profpect that, through your affiftance, we fhall foon become a wholefome fhip.

—— *December* 15.

Since my laft of the 11th inft. I have received eighteen patients with the fever, none of which have died, although fome of them were brought to the hofpital in a ftate of the utmoft danger. The utility of the fumigation appears now very evident, as, notwithftanding the great number of fever patients brought into this hofpital fhip, not one of the attendants, or fhip's company, have experienced the flighteft indifpofition fince we firft began to employ it.—A very fatisfactory demonftration of its power in deftroying contagion ; indeed, Sir, I moft fincerely congratulate you on the fuccefs of a difcovery, which promifes to be of fuch eminent fervice to fociety. Believe me, every thing fhall be done, on my part, agreeably to your directions, to give it its full effect.

—— *December* 19.

The fumigation continues to demonftrate its efficacy, as all the attendants, and fhip's company, continue to enjoy perfect health, notwithftanding I have lately received fome patients with the contagious fever in as bad a ftate as any I have feen ; nor has a new nurfe, or any of the workmen, who are daily employed in the hofpital making the propofed alteration of the neceffaries, fuffered the flighteft attack of the difeafe.—Mr. Menzies goes on with the Ruffian fhips, from which I hope foon to find the infection totally extinguifhed.

—— *December* 21.

—— *December* 21.

I am happy to inform you, that the contagion on board this ſhip appears to be nearly at an end, no one either of the attendants on the ſick, or of the ſhip's company, having been attacked with the fever ſince we began to fumigate, notwithſtanding we have received ſome patients in as bad a ſtate of fever, ſince that time, as any from the firſt importation of the diſeaſe. The people bear it onceedingly well, and I frequently ſtand in the midſt of a cloud, ariſing from the fumigation, as thick as a fog, without the ſmalleſt inconvenience, a circumſtance of great conſequence, as the ſick are all in the wards during the fumigation, and their clothes, &c. are conſequently impregnated with the acid vapour. In a few days we ſhall be able to aſcertain the ſucceſs on board the Pamet Euſtaphia, the only ſhip at this port in which the fever at preſent ſeems to prevail to any degree. I ſhall then conſider the experiment as complete, and ſhall congratulate you on the ſucceſs of an invention, that, in all probability, will give you immortal honor, and which, from its public utility, you will ſo highly merit.

—— *December* 30.

Since my laſt, one nurſe and one marine have been taken ill of the fever, although the ſymptoms are evidently milder than heretofore. As it is impoſſible to ſay how long contagion may remain in an infected perſon before it is put into action, I am not at all diſcouraged by theſe two caſes; but ſhall continue every exertion in my power, in proſecuting the experiment, which has already been of ſuch eminent utility. I have received ſeveral patients from the St. Alexander Niewſki, and another Ruſſian ſhip, returned from ſea, five or ſix of them ill

I   of

of the fever. I need hardly obferve, that if you could by any means enforce the fumigating all the Ruffian fhips, as well as enjoin cleanlinefs, it would be of the utmoft confequence; and if you can fend any perfon to aid me in the bufinefs, I fhould be very glad; for though Mr. Menzies fo ftrongly re-commended, to the Commanders, the neceffity of continuing the fumigation, not one at Sheernefs has made application to me for any materials for that purpofe.

Extract of a Letter from Mr. Baffan to Mr. Menzies.

Sheernefs, December 30.

The Ruffian fhips which arrived this week from fea, are fickly. I received feveral with the petæchial fever, as bad as any I have feen; and am forry to fay that nurfe Murray has had a flight attack, and one of the marines is at prefent ill of the fame fever. He was taken ill on Tuefday morning, the fymptoms are not fo violent as formerly, and I fhould double my attention in profecuting the experiment which has already been of fo much ufe.

TO DR. CARMICHAEL SMYTH.

Sheernefs, January 4.

I cannot account for the contagion having produced the effects on the people mentioned in my laft

laft, otherwife than from the fumigation having been ufed the preceding week, only once a day, or from their having been infected prior to the commencement of it, which I think is not impoffible. I am now determined to ufe it conftantly twice a day, and have done fo fince Tuefday laft, the day on w'ich the marine was attacked; befides, excluſive of the general fumigation, I place a fumigating pot or two in the wards near the worft of the fever patients. The fick not only bear the fumigation exceedingly well, but aid us voluntarily every day, the convalefcents carrying the pipkins about, and expreffing their conviction of its keeping the wards sweet, which certainly it does, and thofe perfons who have hitherto efcaped infection, are fo much convinced of its efficacy, and have fo much faith in its power, that I fhould find it difficult to difcontinue the ufe of it, whilft there is a fick man on board. As a week has now elapfed fince any perfon has been attacked with the difeafe, notwithftanding we daily receive patients in the fame putrid petæchial fever, from the fhips lately arrived from fea, I have every reafon to expect our being once more a wholefome fhip. Be affured, Sir, that no pains fhall be fpared, on my part, to accomplifh fo defirable an object.

—— *January* 7.

I am happy to inform you that no perfon has been attacked with the putrid fever fince my laft, though we have received feveral fick from the Ruffian fhips lately arrived from fea.

—— *January* 13.

I am happy to acquaint you, that fince we began again to fumigate the fhip twice a day, no one has
been

been attacked with the fever, although there are
feveral carpenters at work in altering the neceffaries,
which are nearly completed on the lower gun deck,
and are to be altered immediately on the other
deck.

I faw Captain Senevin, Commander of the
Pamet Euftaphia, the day before yefterday, who
informed me that he had continued the fumigation
every day fince Mr. Menzies's departure, and that
he had now no fick on board.

—— *February* 3.

I have the pleafure to inform you, that the contagion feems now to be totally extinct, no one having been attacked with the fever fince the 26th of
December laft, and only two fince the 26th of
November, the time when the fumigation was begun; one of thefe a marine, who, ten days previous to his being taken ill, had conftantly drank
very hard, and was often drunk; the other a nurfe,
who was very flightly attacked, and both, in my
opinion, might have received the infection long before it was put into action, as from their duty they
were conftantly expofed to the contagion when it
was firft brought into the fhip; and this is rendered
ftill more probable, as there have been feveral arti-
ficers at work, making the alteration in the privies,
and of courfe amongft the fick, and likewife a frefh
nurfe, a young woman immediately employed in
the fever ward, none of whom have received the
fmalleft injury. I therefore now confider the experiment as complete, and can bring fufficient evidence
to convince any one that the contagion in the hofpital, on board the Union, has, through Divine pro-
vidence, been deftroyed by the fumigation you recommended:

commended : befides, as the acid vapour keeps the
fhip fweet, it is my intention to continue it for that
purpofe conftantly, if I am permitted fo to do.
The fick bear it perfectly well, and, from its power
in deftroying alkaline vapour, it renders the air
pure, and confequently. grateful both to the fick
and convalefcents, as well as to thofe whofe duty it
is to attend them. I moft fincerely congratulate
you on the fuccefs of this bufinefs.

And am,

Sir, &c.

A. Bassan.

To the preceding account of the experiment on
board the Union, I fhall take the liberty, my Lord,
to fubjoin a brief defcription of the trials made, at
the requeft of the Ruffian admiral, and with the
approbation of your Lordfhip, on board fome fhips
of that fquadron : and here I muft again refer you
to Mr. Menzies's journal.

REPORT

REPORT

OF THE

EXPERIMENT

FOR

Stopping the Progress of Contagion, *as executed on board some of the Russian Men of War, by* Mr. Arch. Menzies.

SOON after my arrival at Sheerness, I had the honor of being introduced to his Excellency Admiral Hannikow, Commander of the Russian squadron at that port, on which occasion he was pleased to express a particular desire of having the most sickly ships of his squadron purified by the same process of fumigation, as I was then carrying on, on board the Union hospital ship. This being made known to the Lords Commissioners of the Admiralty, they were pleased to declare their approbation, by requesting Admiral Buckner to confer with his Excellency on this subject: and on the twenty-eighth

of

of November, it was agreed between thefe Commanders, that the fumigation fhould be tried, under
my directions, on board fuch of the Ruffian veffels
as were then moft infected with the contagious fever, which had already proved fo fatal to many of
their crews; and it is but juftice to fay, that his
Excellency, on this occafion, fhewed a particular
zeal for its fuccefs, by offering me every aid and
affiftance, and by affuring me of a ready compliance, with every means that might be fuggefted
to accomplifh fo defirable an object, as the health
and prefervation of thofe under his command. But
it fo happened, that, on the day following, he was
ordered, with part of his fquadron, to the North
Seas, and in this ftate of hurry, not having time
to confider which was the moft fickly veffel, he
left orders for the trial to be made on board the
Revel frigate; but on examining the hofpital books
on board the Union, the Pamet Euftaphia, of 74
guns, appeared to claim our firft attention, from
her fickly ftate: I therefore waited on Admiral
Buckner, to acquaint him with this circumftance,
and he very readily applied to Capt. Chechagoff,
on whom the command of the remaining part of
the Ruffian fquadron had devolved, and obtained
his leave for the trial to be made on board of her,
in preference to the other. After this, fome unavoidable delay was occafioned, in waiting for the
materials, and collecting together the utenfils neceffary for the operation.

In the forenoon of the feventh of December,
1795, I went on board the Pamet Euftaphia, and
having ordered the ports, fcuttles, and hatchways
to be clofe fhut up, with the fhip's company between decks, we fumigated her for the firft time,
and continued it morning and evening on the following

lowing day, in the fame manner that we had done the Union hofpital fhip.

This fhip has of late fent more fick with the malignant fever to the hofpital, than all the reft taken together, of the Ruffian fquadron laying at this port, which her Commander, Capt. Sinavin, attributes in a great meafure to her fhingle ballaft being chiefly compofed of fand, intermixed with a large portion of wet earth, that keeps up a conftant moifture and dampnefs below, in fpite of every means of ventilation : add to this the putrid ftench, arifing in fo clofe and crouded a fituation from the *fhubs* or fheep-fkin great-coats, which are generally worn by the Ruffian feamen, with the woolly fide next their body, and which undoubtedly muft aid to nourifh the feeds of contagion, and increafe its virulence.

I reprefented to feveral of the Commanders of the Ruffian men of war, the neceffity there was of deftroying, or at leaft of fuppreffing thefe *fhubs* in this country, for though they might be very comfortable, and anfwer pretty well in dry, cold, frofty weather, such as is generally the cafe in long winters in Ruffia, yet they were by no means calculated for the chilly wet weather which generally prevailed in this country; as in a damp ftate they never fail to impregnate the air with offenfive putrid effluvia, that muft be extremely hurtful to people's conftitution where it is conftantly breathed by fo many crowded together in fuch a confined fituation.

Early on the morning of the ninth, the Pamet Euftaphia, with the Ratvezan of 66 guns, were removed up to Chatham, in confequence of which it was not in my power to continue the fumigation, though I went there on purpofe. And on the following day, the crew was fo bufily occupied, in unrigging the fhip, and clearing her of ftores and provifion,

provifion, to prepare her for going into dock, that
no time could be fpared to attend to the fumigation,
until that duty was accomplifhed ; which as it would
take up fome days, and as her people were then to
be put on board a receiving fhip, while fhe was in
dock, (a circumftance I confidered as very unfavourable to the experiment) I therefore came to London
on the eleventh, to confult with Doctor Carmichael
Smyth, what plan was beft to purfue ; for as this
was the only Ruffian veffel from which a fair efti-
mate could be drawn of the utility and efficacy of
the fumigation, I was anxious to continue it, in
whatever manner might be thought moft likely to
fecure fuccefs in deftroying the contagion, or leffening its malignity.

I returned to Chatham again on the fourteenth,
with orders to fumigate as many of the Ruffian
veffels, efpecially fuch as were moft fickly, as I poffibly could ; for though the experiment could not
be regularly carried on, yet in this manner it might
leffen the virulence of the diforder, and diminifh the
number of fick fent to the hofpital.

Next day I waited on Captain Chechagoff and
Capt. Sinavin, and found that their veffels were not
yet cleared of their ftores, &c. fo that I could not
go on with either. Indeed, the Ratvezan was
pretty healthy, her Commander, Cap. Chechagoff,
being very attentive to every means of purifying his
veffel by ventilation and cleanlinefs, and by deftroying and fuppreffing the *fhubs*, as far as he poffibly
could ; for he told me, he could not do them away
altogether, without giving the men other clothing
in lieu, which muft be done by an order from the
Commander in Chief.

Captain Chechagoff alfo informed me, that the
Pimcn, of 66 guns, was arrived at Sheernefs, which

K   had

had fome time ago been fo very fickly, that boats
from other veffels were forbid coming along-fide of
her, from a dread of the infection; he therefore
expreffed his defire of having her well fumigated;
and I immediately fet out to execute his requeft.

On the fixteenth of December, I fumigated the
Pimen for the firft time. Her crew, however, was
nowife fickly now, although on vifiting her between
deck, before the fumigation, the ftench produced
by the fhubs was very perceptible, and extremely
offenfive; and it was pleafing to obferve the fudden
change produced by the powers of the nitrous vapour in deftroying it.

When I went on board, on the following day,
to continue the fumigation, I found the officers and
crew attending Divine Service, and the Priefts
fprinkling the decks with Holy-water, fo I did not
intrude; but left orders with their own furgeon,
to fumigate the fhip in the evening, if he could con-
veniently, which he did. As this was a holiday
amongft them, I alfo declined calling on board the
Revel frigate till the next day, when, after fumigating the Pimen, I went on board the Revel, to requeft them to prepare for fumigating her. On
vifiting this veffel between decks, I found the putrid ftench from the fhubs extremely offenfive and
difagreeable from the confined air, and want of
ventilation; and I had great difficulty to make myfelf underftood, or give any particular directions
for want of an interpreter.

The nineteenth was fo boifterous that I could
not get on board either veffel, but the Surgeon
of the Pimen was fo good as to continue the fumigation as ufual. Captain Colokolfoff, the Commander of this veffel, was extremely civil, and well
difpofed to promote my endeavours; and the principal

pal officers were equally polite and ready to fee my directions executed on all occasions.

Next day I visited the Pimen, which was now quite free from stench or any offensive smell, in consequence of the fumigation having been regularly continued. I also began to fumigate the Revel frigate, and regularly attended both vessels, for the three following days; after which I left the materials and utensils on board them, with directions to their own Surgeons to continue it in the same way daily.

From what information I could collect, the Revel had not been very sickly, yet the few she had lately sent to the hospital, were malignant fevers, which clearly shewed that the contagion was lurking on board her, though it did not spread with much violence.

Being particularly anxious to resume the experiment on board the Pamet Euftaphia, she being the most sickly, and on that account claiming more particular attention, I came up to Chatham on the twenty-fourth, and found she had been just hauled into dock, and her crew put on board the Prince Edward receiving ship, where they were very much crouded. On the following day I began the fumigation, but as many of the ports and hatchways of the ship could not be shut close enough to retain the vapour for a sufficient length of time, a quick and strong fumigation became more essentially necessary; which, however, I could not get them to execute, not being able to make them understand my meaning, for want of a sufficient knowledge of their language.

The fumigation was, notwithstanding, continued regularly on board this ship for the four following days, although it was not in my power to prevail

on

on them to do it fufficiently ftrong, to do juftice to the trial,—and to infure that fuccefs we had already experienced on board the Union ; their excufe generally was, that the fire was too much occupied to get a fufficient quantity of fand heated.

But, as it was poffible, that even this flight fumigation might fucceed by long continuance, and as their own Surgeon was now acquainted with the procefs, and well difpofed to carry it on, I left the materials and utenfils on board ; and, before my departure, waited on Captain Sinavin, who, at this time, lived on fhore, and who (after being acquainted with the foregoing circumftances) faid that he fhould order it to be continued while his fhip's company were anywife fickly.

The Ratvezan having likewife gone into dock ; to prevent her crew becoming fickly on board the receiving fhip, I, at Captain Chechagoff's particular requeft, fent utenfils and materials on board to fumigate daily.

Having now put thefe Ruffian veffels in a fair train for continuing the fumigation ; and finding that my prefence, on account of my ignorance of their language, could not be of any further fervice ; at the fame time, fome urgent bufinefs, of our late voyage, preffing hard upon me, I returned to town on the thirtieth of December, leaving the further profecution of the experiment, as above related, to be conducted by their own Surgeons ; and I have the moft pleafing hopes that it will be attended with beneficial effects to her Imperial Majefty's fubjects, not only in the prefent inftance, but in every fimilar fituation hereafter.

ARCHIBALD MENZIES.

Having

Having now, my Lord, finished with the account of the experiment given by the two gentlemen who have been employed in conducting it, permit me, before concluding the subject, to call for a moment your Lordship's attention to some of the principal circumstances, and to the conclusions which they afford.

In the first place, my Lord, it must be allowed · that the present experiment fully justifies all I have said respecting the safety with which the nitrous acid (procured in the manner described) may be employed as a fumigation. No one surely can say that I assume too much, when I consider the safety of the fumigation as established, after a trial of nearly three months, for an hour and a half or two hours, morning and evening, each day, on board an hospital ship, containing from two to three hundred persons of different sexes, and ages, and labouring under different diseases; without a single instance of permanent inconvenience or bad consequence arising from it: for the slight cough, which it at first excited, and which was evidently owing to the aukwardness and ignorance of those who carried the fumigating pipkins, cannot be looked upon as such, and no farther inconvenience has ever been felt by any one on board.

Having established then this important fact, that the nitrous acid is attended with no risk to the health or safety of the people exposed to it, let me next claim your Lordship's attention to the sensible and immediate effects of it.

We are told by Mr. Menzies, that after the first fumigation, and still more remarkably after the second, the air of the hospital was perceived to be purer, and free from any putrid or offensive smell; these immediate effects of the fumigation, are likewise wife repeatedly mentioned by Mr. Baffan, the laſt of them indeed was too ſtriking not to be taken notice of by every perſon on board. That the va-pour of the nitrous acid ſhould be found to deſtroy an offenſive ſmell, the effect of animal exhalations, I was not ſurpriſed at, having myſelf had repeated experience of the fact; but that it would alſo ren-der the air purer and more proper for reſpiration, I was by no means certain, until I found the repeat-ed obſervations of thoſe Gentlemen, confirmed by the evidence of Mr. Keir, of Birmingham, one of the firſt chemiſts in this country, or perhaps in Europe; an extract of a letter from this gentleman, whom I have not the honor to know perſonally, to a friend of his in town, I have ſubjoined for your Lordſhip's ſatisfaction, as it affords a convincing proof, from chemiſtry, of the truth of what Mr. Menzies and Mr. Baffan obſerved in practice. Theſe two qualities, my Lord, viz. the rendering the atmoſpheric air purer, and conſequently fitter for the purpoſes of animal life, and the completely deſtroying the offenſive ſmell reſulting from animal effluvia or putrid matter, are, of themſelves, con-ſiderable advantages, if no others were to be ex-pected or derived from the fumigation; but they are of ſtill higher importance, when conſidered as preſumptive evidence of the power of the nitrous vapour to deſtroy contagion; for whatever is found to deſtroy the ſmell of putridity, and at the ſame time to render the air purer, we muſt ſuppoſe more or leſs conducive to this grand object. But pre-ſumptive evidence, on a ſubject of this importance, in which the lives of thouſands are involved, is not ſufficient to ſatisfy the mind; and happily for man-kind, the preſent experiment, inſtituted under your

Lordſhip's

Lordfhip's aufpices, affords complete and direct evidence of the fact.

But to bring this home to the underftanding and conviction of all mankind, it is only neceffary to look with attention, at the annexed Hofpital Return; for by comparing the ftate of health of the fhip's company, with the progrefs and effects of the contagion, before and after the experiment was begun, a clear and decided judgment may be formed of its effects, even by the moft ignorant. They will in the firft place obferve, that from the 3d of September (the day the Ruffians, ill of the fever, were firft brought on board) to the firft of October, there were nine perfons feized with the diftemper, one of whom only belonged to the fhip's company; the others were attendants on the fick. That in the month of October, eight perfons more were attacked with the difeafe, and of thofe three belonged to the fhip's company. But, that from the firft of November, to the 26th of that month, twelve perfons were attacked with the difeafe, among whom we find eight belonging to the fhip's company. From this fhort ftatement it is evident, that the contagion which was at firft chiefly confined to the hofpital, affecting thofe only who were immediately employed about the fick, had gradually fpread over the fhip, and been communicated to the fhip's company; by which means the ficknefs and mortality had increafed : and the probability is, that had not a ftop been put to it, it would have gone on increafing in proportion to the diffufion of the contagion, and to the increafing defpondency of the people, who confidered themfelves as fo many devoted victims. The whole number of perfons feized with the diftemper, during the firft three months that it prevailed on board the fhip,

was

was thirty (befides fix children) which was more than one-third of all the people in the fhip, who were only eighty-five, officers included. Of the thirty feized with the fever, eight died of the immediate effects of it, a large proportion furely, being not much lefs than one in three, and which fufficiently marks the malignity of the diftemper.

Having taken a view of the ftate of the fhip's company, and of the progrefs of the contagion before the experiment, let us now, my Lord, turn to the other fide of the picture, and fee what was the fituation of things after the fumigation was begun.

On the 26th of November, the fhip was fumigated for the firft time, and from that day to the 25th of December, not a perfon on board was attacked with the fever, their defpondency was now changed into joy, and their fear into confidence; but as very great confidence is always dangerous, it proved fo in the prefent inftance. On the 17th of December, they imagined themfelves fo fecure, that they difcontinued the cuftom of fumigating the fhip morning and evening, thinking that once a day was fufficient; the trial, perhaps, was worth hazarding, but on the 25th of December, one of the nurfes fuffered a flight attack, and on the 26th, a marine, who for a week preceding had been in a ftate of intoxication, was feized with the fever, of which he died. Thefe two accidents gave immediate alarm; they returned again to the practice of fumigating twice a day, and from that time to the end of the diforder, there has not been an inftance of a perfon fuffering from contagion on board the fhip. But the advantage of the fumigation was not felt by the fhip's company and attendants alone, whom it preferved from the baneful effects of the fever, the fick and convalefcents derived almoft an

equal

equal benefit from it. The fymptoms of the difeafe (as Mr. Baffan expreffes it) were meliorated, and loft much of their malignant appearance, and the advantage of a pure air, and free from ftench, to convalefcents, may readily be conceived.

From the above relation, my Lord, it plainly appears, that whilft the practice of fumigating the fhip twice in the twenty-four hours was continued, there was no fymptom of contagion or of difeafe, and that the only two accidents which happened from the commencement of the experiment, to the prefent hour, occurred on the 25th and 26th of December, nine or ten days after they had ceafed to fumigate the fhip, in the manner I had directed. The attack of the nurfe, indeed, was but trifling, and I think it not improbable that the fever, as well as the death of the marine, were the confequence of his own intemperance; at any rate fuppofing both the one and the other to have fuffered from contagion, thefe cafes do not in the leaft invalidate the general fuccefs of the experiment, and only prove, that in a fituation where contagion is conftantly generated, it requires to be as conftantly deftroyed; otherwife it is ready at every inftant, like the hydra, to rear again its peftilential head.

But, my Lord, the fuccefs of the experiment has not been confined to the Union, the power of the nitrous vapour to deftroy contagion, has been equally difplayed on board thofe Ruffian veffels where it has been employed.

Your Lordfhip muft have obferved, in Mr. Menzies's Journal, the many unexpected delays he met with in the execution of this bufinefs. The fudden departure of the Ruffian Admiral, with a confiderable part of the fleet, before the fhips, the moft

L   proper

proper for the experiment, were fixed upon ; Mr. Menzies beginning, in confequence of not being properly informed, with fhips where the fumigation was not fo immediately neceffary ; afterwards when he began to fumigate the Pamet Euftaphia, which had fent more fick to the hofpital than any fhip of the fleet, fhe was immediately ordered into dock, and the crew turned over into a receiving fhip, a fituation extremely difadvantageous for fuch an experiment : not to mention the various difficulties and obftacles arifing from the difference of language, ufages, religious ceremonies, &c. fufficient to have difcouraged a man of a lefs firm mind, or who was lefs zealous than Mr. Menzies in purfuing his object. He perfevered, however, for fome time, but at laft was under the neceffity of returning to town, and of leaving the farther profecution of this bufinefs to the Ruffians themfelves ; and yet, my Lord, owing to the good fenfe and proper conduct of their officers, who, convinced of the advantage of the fumigation, continued the daily practice of it ; thofe fhips that have been fumigated, are free from contagion, and particularly the Pamet Euftaphia which was the moft fickly, is now one of the healthieft of the fleet, and has no appearance of contagion on board, nor a man ill of the fever ; and fo great is the opinion entertained by Admiral Hannicoff, of the efficacy of the fumigation, that he lately fent to town for materials for fumigating fome more fhips.

Such, my Lord, has been the refult of an experiment, by which fome lives have been already faved, and from which two important facts are clearly eftablifhed, viz. the power of the nitrous acid to deftroy contagion ; and the fafety with which it may

be

be employed in any fituation, without inconvenience or ri'k of fire.

It would be, perhaps, improper in me to detain
your Lordfhip any longer on this fubject, by en-
deavouring to point out the importance and extenfive application of the prefent difcovery; a difcovery equally applicable to every fpecies of putrid contagion, even to the plague itfelf; a difcovery therefore, in which all nations are more or lefs interefted,
but whofe utility muft be moft fenfibly felt by our
own; where a commerce, extended to every quarter of the globe, covers the fea with our fhips,
whilft our gallant navy ftill maintains the decided
empire of it.

Oh fortunatos nimium, fua fi bona norint,
Britannos!

I have the honor to be,

My Lord,

With the higheft refpect,

Your Lordfhip's

Charlotte-ftreet,   Moft obedient and obliged
Bloomfbury,
Mar. 12, 1796.   Humble Servant,

James Carmichael Smyth.

Earl Spencer.

A RE-

A RETURN of thofe perfons, amongft the attendants on the Hofpital or belonging to the Ship's Company of the Union, who were attacked with the Contagious Fever, from the 3d of September, 1795, when the Ruffian fick were firft brought on board, to the 10th of February, 1796; the date of the laft report.

(Signed) A. BASSAN, Surgeon of the Ship.

Before the Ship was fumigated.

Names	Quality	When feized	Recovered	Dead
S. Brown	Nurfe	Sept. 6	———	
H. Warren	———	—— 7	———	
M. Mitchel	———	—— 9	———	
M. Reed	———	—— 11	———	
Mr. J. Gardner*	Sˢ 1ft Mate	—— 15		Sep. 24
M. Rawlins	Nurfe	—— 18	———	
S. Hayes	———	—— 20	———	
Tho. Mitchel	Helper	—— 22	———	

* He was difcharged from the Union, and entered on board the Sandwich, the 12th of September; was taken ill a few days after and died in about a week.

Names	Quality	When seized	Recovered	Dead
A. Clavering	Nurse	Sept. 24		Sep. 28
Tho. Lee	Marine	—— 29		Oct. 1
M. Sawer	Washer-wo.	Oct. . 6		—— 15
Mr Messersmidt	Ss 1st Mate	—— 6	——————	-
A. Bright	Nurse	—— 8		—— 11
D. Bower	Ab.	—— 8	——————	
H. Tuberville	Nurse	—— 14	——————	
Mr. Bodker	2d Ss Mate	—— 22	——————	
Cha. Walton	Ab.	—— 22	——————	
James Potter	Marine	—— 22	——————	
C. Taylor	Nurse	Nov. 2	——————	
S. Parker	Washer-wo.	—— 4	——————	
Wm. Crasby	Marine	—— 4	——————	
Wm. Welch	————	—— 10	——————	
Rd. Welch	Ab.	—— 10	——————	
Henry Kelly	—	—— 17	——————	
Peter Parker	—	—— 17	——————	
Geo. Mantle	Marines	—— 18	——————	
Tho. Reed	St Marines	—— 18	——————	
Jos. Copeland	Ab.	—— 20		Dec. 4
Ja. Tuberville	Marine	—— 20		Nov. 24
M. Clay	Washer-wo.	—— 24		T. uncertain

Before the Experiment.——Total   30   22   8

After the Ship was fumigated.

Names	Quality	When seized	Recovered	Dead
Marg. Murray	Nurfe	Dec. 25	{ Recov. in { a few days	
James Farmer	Marine	—— 26		Jan. 6

Since the Experiment.——Total   2   1   1

N. B. On the 25th of November, the ship was fumigated for the first time, and the fumigation repeated twice a day till the 17th of December; from that time to the 26th of December, only once; but from the 26th of December to the 10th of February, twice a day, as at first.

A Weekly

A Weekly Return of the Ruffians received on Board his Majefty's Hofpital Ship Union, in the Malignant Fever from the 3d of September, 1795, to the 28th of January, 1796, exclufive of thofe received in a ftate of debility after the faid Fever, and with other difeafes. By A. BASSAN, Surgeon of his Majefty's Ship Union.

Sept. 1795.	Received.	Difcharged.	Dead.
Sept. 3	37		
—— 10	37	1	
—— 17	4	7	
—— 24	34	7	1
Oct. 1	17	17	5
—— 8	29	15	
—— 15	20	5	2
—— 22	15	14	1
—— 29	18	11	1
Nov. 5	31	9	1
—— 12	21	13	
—— 19	20	44	5
*—— 26	29	39	1
Dec. 3	12	5	1
—— 10	12	16	1
—— 17	35	48	
—— 24	8	25	1
—— 31	40	1	
1796.			
Jan. 7	32	25	3
—— 14	13	7	2
—— 21	20	24	3
—— 28	22	23	6
Total 479		356	34

* This day the Ship was fumigated.

From

From the above return it appears, that the number of perfons ill of the contagious fever, brought on board the Union, the two laft months, December 1795 and January 1796, were nearly equal to the number received the two preceding months, October and November. It alfo appears that for the firft month, after the fhip was fumigated, there were few fever patients who died. The increafe in the number of deaths in the following month, may fairly be afcribed to the return of the fleet.

N. B. The greater part of the Ruffian fquadron failed on a cruize November 29, and returned into port December 27, two or three fhips at a time.

APPENDIX.

APPENDIX.

Extract of a Letter from Mr. Keir, of Birmingham, to a Friend in Town.

January 25, 1796.

I CONSIDER Dr. Carmichael Smyth's difcovery to be very valuable. The fumes in his procefs are quite different from the ordinary nitrous vapour in the diftillation of aqua fortis, or from that which exhales in the folution of metals, by nitrous acid; the latter is highly fuffocating and noxious, and may be called the phlogifticated nitrous acid vapour. The fumes made in Dr. Smyth's manner (if there is no metal employed in the veffel, &c.) is highly dephlogifticated or oxygenated nitrous vapour, and is alfo mixed with a large quantity of pure dephlogifticated air, which is extricated from the materials, and thefe fumes are not only not fuffocating, but have a very pleafant fmell. If the diftinction is not made between thefe two kinds of vapour, it is to be feared that fome perfon, by accident, or in expectation of getting nitrous vapour more expeditioufly, may ufe the metal veffels, or diffolve metals in nitrous acid.

Extract

Extract of another Letter from Mr. Keir, dated near Birmingham, March 3, 1796.

The difference between the white nitrous acid, (called by Dr. Prieftley, dephlogifticated acid, and by the French chemifts, acide nitrique;) and the red acid, called phlogifticated, or acide nitreux, is well known, and was firft particularly noticed by Scheele, who fhews how the one may be feparated from the other by diftillation. There is the fame difference in the colour of the vapours from thefe two acids; and Dr. C. Smyth has himfelf obferved, that the vapours, in the diftillation of nitrous acid, were not noxious; which obfervation he has very happily and ufefully applied. In diftilling the nitrous acid from very fmall quantities of nitre and oil of vitriol, in glafs veffels, and when the materials are very pure, I have feen nothing but the white vapours, fuch as arife in Dr. C. Smyth's procefs, but Scheele fays, that at the end of the operation, fome red vapours, rife, and it may be the cafe when a very ftrong heat is applied. But the very noxious red fumes which appear in the ufual procefs of diftilling aqua fortis, are occafioned, as you mention, by the iron veffels; and the manufacturers even put in old nails and fmall pieces of iron into their pots, in order to give a high degree of red fmoking quality to the acid. When you acquainted me of Dr. C. Smyth's difcovery, it occurred to me, that as the common notion of nitrous acid vapours, is confined to thofe that are red, fome people might, in the firft place, be prejudiced againft it, from the idea of the vapours being noxious; as the red vapours are undoubtedly, and others might think that they made the procefs more effectual, by adding to Dr. C. Smyth's mixture, metals, or inflammable fub-

ftances,

ftances, in order to produce thofe red vapours. I therefore thought it would be proper for Dr. C. Smyth, to point out the difference between the vapours produced in his method, and the red nitrous fumes which are fo well known ; and alfo to caution the operators to avoid metal veffels, or the addition of metals or inflammable fubftances.

There is a good deal of vital air extricated from the mixture, but I cannot agree with thofe who attribute the medicinal effect to it, we know little of this fubject ; but the analogy of the deftruction of all animal and vegetable fermentation by mineral acids, which is well afcertained, inclines me to believe the agency of the acid, in the deftruction of the contagion. The matter of which is, I prefume, animal, in fome vicious kind of fermentation.

A Letter from Mr. Buffan, Surgeon of his Majesty's Ship Union, to Dr. Carmichael Smyth, under date, the 16th of February, 1796.

DEAR SIR.

WE had an increafed mortality amongft the Ruffians laft month, but, thank God, not from the contagious fever, that being now totally extinct ; but from fome being brought in a dying ftate, others in the fcurvy, the moft deplorable cafes I ever faw, added to which, feveral hectic

patients,

patients, who had been declining fome time, happened to die at that particular period. I hope we
fhall have no return of fo dreadful a calamity.

I remain,

Sir,

Your moft obedient Servant, &c.

(Signed)   A. Bassan.

---

*Extract of a Letter to Mr. Menzies, from Captain
Chechagoff, Commanding Officer of the Ruffian
Fleet, in the abfence of his Excellency Admiral
Hannicoff. Dated Chatham, March 9, 1796.*

Agreeably to your wifh, it is with the utmoft
pleafure that I expofe the proofs of a truth fo ufeful for the human kind, and fo much to the honor
of thofe that are the primitive caufe of it, and thofe
that put them in execution, with an efficacy, as is
acknowledged in the certificate here joined. I beg
to prefent my compliments to Dr. Smyth, for
whom I have the refpect that is owing to all thofe
who have enfured their renommée,* by the good
they have done to the public, and to get its fuf-
frage. I am, with much efteem, &c.

(Signed)   P. Chechagoff.

C E R-

* The public will recollect that Captain Chechagoff is a
foreign officer, writing Englifh, and therefore will not be
furprifed at his making ufe of one French expreffion.

CERTIFICATE.

" It has been obferved that the fumigation, with
" the nitrous acid, introduced by Mr. Menzies on
" board the fhip Pamet Euftaphia, has produced, in
" a fhort time, the beft effect in ftopping the pro-
" grefs of the fever and other evils, which were
" then evidently increafing, for which reafon it was
" not only regularly continued on board of that
" fhip, even after Mr. Menzies's departure, but
" adopted on board of others, and always found
" ufeful. It is therefore my duty to certify by this
" not only the good confequences that have been
" obferved from that ufeful contrivance, but even
" the advantage that arifes from its eafy and fure
" execution, in comparifon with other means of fu-
" migating the fhips which requires greater atten-
" tion from the fire that muft be made ufe of, and
" therefore cannot be effectuated in all the parts of
" the fhip."

(Signed) CHECHACOFF,

Captain and fenior Officer of the

Ruffian Fleet.

March 10, 1796.

COPY

COPY

OF A

LETTER

FROM

DAVID PATTERSON, ESQ.

SURGEON IN HIS MAJESTY'S NAVY,

AND LATE

SURGEON TO THE PRISONERS OF WAR AT FORTON;

TO THE

COMMISSIONERS

For taking Care of Sick and Wounded Seamen.

Copy of a Letter, &c.

Gentlemen,

LAST winter, while I had charge of Forton hofpital, although in the midft of very fatiguing duty, and engaged, as I was, in making fome favourite experiments of my own, yet, being extremely anxious to acquire fome practical knowledge of Doctor J. C. Smyth's nitrous vapour, I failed not, after receiving your authority, to put his fumigating plan in execution, as extenfively as it was poffible; nor, at the fame time, to note down the phænomena, as they occurred, with as much accuracy as my leifure time would permit. And now, in order that you may fee, in the faireft points of view, fome of the effects of the vapour refulting from that highly ingenious, and very falutary procefs; and alfo, with the view no lefs of doing juftice to Doctor J. C. Smyth, than of rendering his nitrous vapour more extenfively ufeful, in the Navy, Army, &c. I beg leave to communicate to you the contents of the following pages; which, fhould they appear to you in any degree calculated to be ferviceable, in promoting the welfare of thefe realms, I humbly requeft you will be pleafed to lay them before the Right Honourable the Lords Commiffioners of Admiralty.

In purfuing my prefent plan, I fhall, after giving you a fhort, but not imperfect account of the method I followed in fumigating the wards of the hofpital, ftate, in a faithful manner, fuch facts, whether of a general, or of a particular nature, as arofe from the trials that were made; after which, by moft humbly offering a few obfervations connected with the fubject, I fhall conclude my letter.

N

Every

Every evening a certain number of wards were fumigated, each by means of three pipkins, for an hour; the gally-pot in each pipkin containing the quantities of pure nitre in powder, and concentrated vitriolic acid as directed by Doctor Smyth.* Three persons, each carrying one fuming pipkin, went round a ward, following one another at some distance, and holding the pipkins under each bed, for a considerable time, as they went along; and they continued doing so as long as the fumigation lasted. The wards, by opening the windows and doors, were afterwards filled with atmospheric air.

The vapour, proceeding from the decomposition of the nitre, by means of the concentrated vitriolic acid, was in such great quantity, that a ward 57 feet by 20, and 10 feet 6 inches high, was filled with it, by means of three pipkins, in the manner I have mentioned, in the space of fifteen minutes.

On the wards being filled with nitrous vapour, some of the patients who laboured under affections of the lungs were seized with fits of coughing; none of them, however, to any great degree. With little or no exception, the patients, in the wards that were fumigated, bore the vapour without feeling any disagreeable effect from it. If, indeed, a pipkin was, accidentally, held very close to the mouths of any of the patients, which, from awkwardness, was sometimes the case, coughing was immediately produced; and, in one instance, vomiting was occasioned. These circumstances, however, did not prevent the patients from becoming, in a short time, very fond

of

* For a more particular history of the process Vide, Doctor J. C. Smyth's letter to the Right Hon. Earl Spencer, &c. &c. &c. containing an account of the experiment made on board the Union hospital ship, to determine the effect of the nitrous acid in destroying contagion, &c.

of the fumigating bufinefs. For my part, I frequently remained in a ward during the whole time of the
fumigation, often indeed with a fuming pipkin in
my hand, without experiencing any difagreeable effect whatever. The fume was to me pleafant. When,
during the fumigation, I remained in a ward, I always wore black clothes, which, even, after being
repeatedly expofed to the nitrous vapour, were not
in the leaft either ftained, or changed from black to
a brown colour.

In the mornings, particularly in dry weather, the
wards that had been fumigated the preceding evening, even although they had been wafhed early in
the morning, and the windows kept open, had a
very agreeable fmell, much more pleafant than that
which was experienced during the fumigation. By
this agreeable odour, in the mornings, I was able
to judge whether or not due pains had been beftowed, the preceding evening, in fumigating the wards.

One dyfentery ward, one fever ward, and one
furgery ward, containing the worft kind of Ulcers,
were, at firft, the places filled every evening with
the nitrous vapour ; but, as the good effects refulting
from the fumigation were to me very obvious, I foon
ufed it more extenfively. The patients, in general,
who laboured under old dyfenteries, many of them
contracted in the Weft Indies, feemed to be greatly
relieved ;* the fevers, which were of no uncommon
genus, and which were in their nature very mild,
foon difappeared, without exhibiting any fymptoms
of typhus ; and the ulcers, inftead of further dege-
nerating or fpreading, put on a favourable appearance, and healed.

It

* Ultimately, a great number of the old dyfenteries, where
the patients were not far advanced in life, did well.

It is, I presume, of no small consequence to observe, that, excepting some marked cases of dysentery, among the servants of the prison and hospital,
in the months of August and September, before the
arrival of the prisoners from the West Indies, and
one case of typhus (in ward 18) with now and then
a case of small-pox, among the West Indians, after
their arrival, there was not any contagious febrile
disorder that made its appearance within the walls
of the hospital, while I had charge of it, notwithstanding the many sources of contagion to which, in
my opinion, all in and about it were exposed. During the last five months of my time, no fewer than
1686 patients were admitted into the hospital, as
may be seen by the hospital books.

The following table serves to shew, at one view,
the highest number of patients in the hospital, the
number discharged and dead, weekly, for four weeks
before, and six weeks after the nitrous vapour was
first used; viz. from the 16th of October, to the
26th of December, 1796. To include a greater
space of time would be improper; because, before
the 16th of October, there were but few patients
in the hospital; and, because, after the 26th of December, there were a great number of extremely
bad cases of gangrenous feet, pneumonia, &c. received into it, from Portchester hospital, and from
the Vigilant and Captivity prison ships.

TABLE.

TABLE.

Before the Nitrous Vapour was ufed				After the Nitrous Vapour was ufed.			
Weeks	Higheft Number in the Hofpital	Number dif- charged	Number dead	Weeks	Higheft Number in the Hofpital	Number dif- charged	Number dead
1	223	2	8	1	340	27	6
2	372	4	21	2	332	7	5
3	371	0	13	3	342	11	8
4	369	1	9	4	340	8	4
				5	486	12	1
				6	539	63	5
		7	51			128	29

After thefe general obfervations on the nitrous vapour, I fhall humbly beg leave to offer the following Cafes, in which it was, undoubtedly, ufed with very remarkable fuccefs.

CASE I.—Jean Louis, French prifoner, of colour, eighteen years of age, from the Weft Indies, was admitted into the hofpital on the 28th of October 1796, for an ill-conditioned Ulcer on the inferior and interior part of the right leg. After he had been fome time in the hofpital, the ulcer began to put on a favourable appearance, and was foon confiderably diminifhed in its fize, merely by means of fimple dreffings.

On the 29th of November, however, the ulcer according to the common phrafe, became foul ; and, by next day, it had fpread to fuch a degree, that it was nearly as extenfive again as it ever had been, attended with very acute pain, and with a very copious thin dark-coloured fetid difcharge. The patient's pulfe, at this time, was 120, tongue clean ; appetite impaired, belly open ; fleep much difturbed.

31ft. From the 29th to this time, a common poultice, thrice a day, was the only application ; but

now,

now, in addition to the poultice, the ulcer was dreffed with the powder of Peruvian bark ; a cooling
medicine, with an opiate at bed-time, was ordered ;
and a vegetable diet with milk was enjoined.

2d December. The ulcer ftill more extenfive than
it was on the 31ft ult. It now extended from the
Tarfus fix or feven inches upwards, and from the
Tibia more than half round the leg : it was ftill in
a floughing ftate, with high, reflected edges. The
other fymptoms much the fame as before. Finding
that the plan hitherto purfued had not produced
any good effect, either on the ulcer or on the fyftem,
the whole of it, excepting the poultice, was aban-
doned, and the nitrous vapour adopted. The ward
in general, and the bed of the patient in particular,
were carefully fumigated, the ward once, the bed
twice a day.

3d. The ulcer had ftopt fpreading, and in fome
places looked clean. Such a fudden change was to
me aftonifhing. Pulfe now 110 ; tongue clean ; belly open ; flept better on the night of the 2d than
for fome time before, notwithftanding the omiffion
of the opiate.

4th. The ulcer was clean, and difcharged good
matter. The patient felt himfelf comfortable. Pulfe
about 90.

6th. The difcharge continued to be good ; and
the ulcer had made confiderable progrefs in healing.
The patient felt himfelf perfectly eafy, and his
health was already very much mended.

The fumigation was continued until the 26th of
December ; from which time, owing to a want of
materials, it was difcontinued until the 11th of January, 1797. The ulcer, during the time the fumigation was ufed, and even to the 1ft of January,
1797, continued to heal kindly, and rapidly ; but,

at

at that period, it again became foul and floughing, and was foon as extenfive as before. The appetite was again impaired; pulfe 120, and fmall; belly open; the patient much weakened and emaciated. Half a drachm of Peruvian bark, thrice a day, and eight ounces of wine in the twenty-four hours, were ordered; and the ulcer was dreffed twice a day with the powder of common Peruvian bark and common poultice. An opiate was occafionally allowed at bedtime. This treatment was preferved until the 11th, when, having experienced no good effects from it, it was difcontinued, and recourfe again had to the nitrous vapour, and common poultice, as on the 2d of December, 1796.

12th. The ulcer had ftopt fpreading, and in fome places had begun to clean. The pulfe was lefs frequent, and more full; and the patient was, in every refpect, better, and more comfortable. The nitrous vapour, &c. were continued.

13th. The ulcer was perfectly clean, with florid granulations, and with about the eighth of an inch of new fkin round the edges. The plan was continued.

20th. The ulcer looked very healthy, and was contracting rapidly. The plan was continued.

5th February. The ulcer had contracted more than one half. The plan was continued.

12th March. The ulcer was nearly healed, (not fo much as the breadth of a fixpence being open) and looking healthy. On this day I finifhed my duty, and, confequently, my obfervations, at Forton hofpital.

I have here to obferve, that about the 1ft of January 1797, all the ulcers in the fame ward (N° 14) with the above, were more or lefs in a bad ftate; and that they all, about one and the fame time, began

gan to put on a favourable appearance; and also
that, in a fhort while, many of them healed. Likewife, it is neceflary to obferve, that particular attention was paid all along to cleanlinefs and ventilation.

CASE II. La Granade, French prifoner, aged
26 years, from the Weft Indies, was admitted into
the hofpital on the 16th of December, 1796, for
chilblains. In the end of February 1797, an ulcer
broke out on his left leg, which became very foul
and floughing, and did not yield to common remedies. On the 7th of the following March, the nitrous vapour was ufed, exactly in the fame manner
as in the preceding Cafe, and by the 12th the ulcer
was perfectly clean.

CASE III.—Elie Double, French prifoner, aged
22 years, from the Weft Indies, was admitted into
the hofpital on the 28th of October 1796, for an
ulcer on the anterior and middle part of his left leg.
By the middle of February, 1797, the ulcer was
cicatrifed, but with a confiderable protuberance
remaining over that part of the tibia, as if the peri-
ofteum and even the bone itfelf, had been in a difeafed ftate. About the end of February the cicatrix became inflamed, foon fuppurated, and degenerated into a foul floughing ulcer, which, inftead of
yielding to any of the various applications, got worfe
and worfe every day. From the end of February,
(I cannot exactly tell the day,) cataplafms of different kinds, myrrh, and Peruvian bark, were tried
externally; and wine, Peruvian bark, opium, &c.
were adminiftered internally. At the fame time,
great attention was beftowed in keeping the ward
extremely clean, and thoroughly ventilated. Finding not only that no good effect was produced by
any of thefe means, but even that the ulcer, the

found

found parts being ftill in a mouldering ftate, grew more and more extenfive, I came to a determination, confidering myfelf fufficiently authorifed, from the experience I had had, to make trial, in this unto-ward cafe, of Dr. J. C. Smyth's nitrous vapour. Accordingly, on the 7th of March the fumigation was put in practice, in the fame way as in the fore-going cafes; and, it is with heart-felt pleafure I re-late it, by the 12th, that was in five days time, and on the day I finifhed my duty at Forton, the ulcer was perfectly clean and healthy.

CASE IV.—Francois a negro French prifoner, age unknown, was admitted into the hofpital on the 26th of January 1797, for a wounded little finger. On examining the wound, I found that the laft bone and the furrounding liguments were the parts moft materially injured. The bone was fractured, and the foft parts were contufed to a very great degree, with a fmall lacerated wound at the tip of the finger. Deeming it neceffary, I immediately amputated the limb at the joint formed by the fecond and laft phalanges. The ftump, the bone being well covered, and the foft parts looking healthy, had all the appearance of doing well, during the firft fortnight; but, unfortunately, at the expiration of that period, it began to put on a very unfavourable afpect. Inftead of the difeafed parts being finuous, or having, what is perfectly underftood in furgical language, a glafly appear-ance, which fometimes indicate a difeafed bone, they became enlarged to a prodigious degree, re-flecting very confiderably, fo as to refemble a ball on the end of the ftump; and, at the fame time, appeared foul, difcharging a dark thin fetid matter. In this ftate, Peruvian bark, opium, &c. were tried, as alfo cataplafms, but to no purpofe. In

O   the

the end of February, recourse was had to the
nitrous vapour ; and, by means of it, in fix days
time, the ulcer was perfectly clean.

CASE V.—Baftern, a negro French prifoner,
age unknown, was admitted into the hofpital on
the 28th of January 1797, for an ulcerated toe.
This cafe very fimilar to the finger of which I have
juft taken notice, and, like it, after various in-
effectual applications, was cleaned, and put, feemingly, in a healthy ftate, by means of the nitrous
vapour ufed according to the manner I have already
related. The four laft cafes were in the fame
ward, No. 4.

Having, with refpect to Dr. J. Smyth's nitrous
vapour, agreeably to my promife, finifhed the moft
important part of my experiments, with the phænomena refulting from them, in order to fhew you that
vapour, under proper management, is capable of
producing very happy effects on the human frame,
I fhall now moft humbly offer a few obfervations
which appear to me to be connected with the fubject.

And, to proceed ; I am in great hopes that the
facts detailed in thefe pages, while they ferve as fo
many proofs of the utility of Dr. J. C. Smyth's fumigating plan, will, at the fame time, anfwer the
happy purpofe of not only removing the ill-ground-
ed fear of Dr. Trotter, and of convincing him, as
well as thofe who think as he does, that no danger is to be apprehended from the combination
of azote with the nitrous vapour ; but, alfo,
of conquering the prejudices of thofe gentlemen
who imagine (for fome highly refpectable medical practitioners have lately mentioned to me
their apprehenfions) that that vapour, from its being loaded with vitriolic acid, muft be intolerable

to

to the lungs, and of courfe highly pernicious to perfons fubjected to its influence.

With regard to the bad, or deleterious effects of
the nitrous vapour, I cannot fay, from experience,
that I am acquainted with any of them. The trials
that I made of that vapour were on a great number of difeafed perfons, who, although crowded
together within the walls of an extenfiva hofpital,
and from that circumftance, as well as others of
at leaft equal moment, expofed to the influence of
noxious effluvia, were obvioufly, in many inftances,
as already mentioned, benefited by its falutary effects. Many patients were cured; others were
put in a fair way of being cured. And, I muft
add, for it is not, I prefume, altogether improbable, that by means of the nitrous vapour, with other
no lefs important meafures, which I adopted, and
inceffantly followed when in my power, the patients who were under my charge, in Forton hofpital, were preferved from the attacks of contagious
fever. I have ventured to fay with other no lefs
important meafures, becaufe I am well aware, as
Dr. J. C. Smyth undoubtedly is, that, without the
moft ftrict attention to cleanlinefs, and to the circulation of pure or atmofpheric air, neither the nitrous vapour, nor any thing adminiftered with
fimilar intentions, can prove fo efficacious as we
could wifh, in preventing or putting a ftop to contagious fever, as well as other difeafes, though
perhaps not fo immediately, yet ultimately as fatal.
I am here under the neceffity of obferving, having
forgot to do it in the proper place, that the folitary
cafe of Typhus, which, as before mentioned, was
in ward eighteen, did not originate in the hofpital,
but in the Captivity prifon fhip. What the nature
of the diforder had been primarily, I am at a lofs

to

to fay. The patient died, and, fortunately, fo did the difeafe; for I faw not another fever of a fimilar nature in the hofpital.

As no contagious fever (I mean typhus, or what fome authors have called jail fever, others hofpital fever, &c.) prevailed, during my time, in Forton hofpital, I cannot fay pofitively that the nitrous va-pour folely prevented fuch a fever from prevailing. All that I can fay is only, that the circumftance of no contagious fever having prevailed in Forton hof-pital, during my time, may be confidered as being of a very fingular nature; more efpecially when we take into our view the vaft number of patients, in the moft filthy ftate, from the Weft Indies, &c. that were received. The very particular attention that was paid to the patients, on their being re-ceived, in ftripping them of all their clothes, in bathing them, in fhaving their heads, in burning all their clothes, and alfo in keeping the hofpital, at all times, extremely clean, and thoroughly ven-tilated, may, it is probable, have contributed not a little towards preventing contagious fever. And, further, another circumftance which, perhaps, had operated very powerfully in affifting to obviate con-tagious fever, and which deferves to be very parti-cularly remarked, was the changing of the wards as frequently as it was poffible, and that according to the nature of the complaints they contained; for inftance, and by comparifon, wards that contained convalefcents, and alfo thofe that contained flight or chronic difeafes, were changed frequently; thofe that contained febrile difeafes more frequently, and thofe that contained very bad furgical cafes moft frequently. By the changing of wards, I mean the removing of the patients from one that they had

had occupied for fome time, to another that was perfectly purified.

When a ward of whatever defcription was changed, it was firft emptied, by the patients being removed into another, and by its bedding being fent to be baked, fumigated, or wafhed; and then it was without lofs of time fumigated by means of fulphur; then white-wafhed; then its cradles cleaned, and wafhed with vinegar; then the floor of it thoroughly cleaned; and, laftly, its windows on the one fide, and its fcuttles on the other, were kept open, always when the weather would permit, until it was again occupied by patients, which, if the ftate of the hofpital admitted, was not before eight days had expired. Such regulations as the above ought, in my humble opinion, to be conftantly and very particularly obferved, by all medical men who have the immediate charge of hofpitais for prifoners of war: many of them, I prefume, might, with propriety and utility, be obferved in any hofpital; and, in concluding this fubjcft, I beg leave humbly to fuggeft to you that no hofpital ought to be full, but, on the contrary, that there fhould always be, in all, according to their different fizes, two, three, four, or more wards left empty, for the very falutary purpofe of changing.

Cleanlinefs, ventilation, and changing of the wards, whether with the view of obviating or removing difeafes, are, in all hofpitals, as well as in all places where prifoners of war are confined, &c. abfolutely neceffary: where they are obferved, medicines will become lefs needful; and when needful, they will, in their operation, be more effectual; but, where they are neglected, the phyficians and furgeons will be fubjected to the very unpleafant trouble of giving their attendance, and of prefcrib-

ing,

ing, to very little purpofe. Warmth ought alfo to be attended to.

In fuch eftablifhments as Forton, cleanlinefs, a free circulation of air, proper diet, &c. ought, agreeably to the very particular orders which you iffue, to be moft rigidly attended to, from one end of thefe eftablifhments to the other. But, I am afraid, orders are not always rigidly executed. The unpardonable neglect of fervants, in not executing, with promptitude and fcrupulous punctuality, the orders with which they are entrufted, is to be lamented, but, I fear, not to be, on all occafions, either prevented or corrected. From what I know of the eftablifhments in queftion, I fhall venture to fay, that, were they always to be properly conducted by the fervants who have the immediate charge of them, we fhould hear lefs frequently of the prifoners, &c. falling a prey to contagious fever than we have hitherto done. This is a fubject, however, with which, at prefent, I fhall not further concern myfelf, excepting to make the following obfervation, which is, that while due care is not taken, in the firft inftance, to prevent contagion from taking effect, the ufe of Dr. J. C. Smyth's nitrous vapour becomes, undoubtedly, the more particularly neceffary : but I am extremely forry to think that Dr. Smyth's plan, as well as others equally well intended, fhould not always be put in execution, but more efpecially in cafes of emergency, with that facility, with that eagernefs, with that candour, which duty, juftice, and humanity, continually require.

Although I have, in the courfe of thefe obfervations, laid very confiderable ftrefs on cleanlinefs, ventilation, changing of wards, &c. yet I would not, by any means, wifh it to be fuppofed that

I have

I have 'done it with the view of fuperfeding the ufe of the nitrous vapour : on the contrary, while on the one hand I am fenfible that the nitrous vapour cannot, without cleanlinefs, ventilation, changing of the wards, &c. be fo efficacious as we could wifh, in putting a ftop to contagious fever ; I am, on the other hand, no lefs fenfible that that fever, when raging to a violent degree, cannot be exterminated by means of cleanlinefs, ventilation, &c. without the affiftance of fome other means. With refpect to hofpitals, fhips, prifons, &c. where people are crowded together, where the introduction of contagious fever is dreaded, or where it actually prevails, the nitrous vapour, with due attention to cleanlinefs, ventilation, &c. may, at once, I prefume, not only be confidered the moft convenient, the moft elegant, and the moft ingenious, but alfo the moft efficacious remedy for the purpofe of counteracting different fpecies of contagion, that has yet been offered to the public.

Further, although cleanlinefs, ventilation, and the changing of wards, very ftrictly attended to, might, in a very great meafure, prevent contagion from taking effect, or from fpreading extenfively, yet, fuppofing them to be attended to as ftrictly as, from the nature of things, it is poffible, they could not, I am too much afraid, deftroy contagion, when prevailing in an extenfive hofpital, &c. For example, let us fuppofe only five or fix hundred patients confined in Forton hofpital, and labouring under contagious fever ; and let us alfo fuppofe it neceffary, for the fake of cleanlinefs, and of putting a ftop to the contagion, to completely fhift all thefe patients once, perhaps many of them twice, and fome of them even thrice, every day ; how, give me leave to afk, would it be poffible to furnifh

with such a great number of patients so frequently
with the clean things required? For my part, I am
fully persuaded that it would prove difficult; so
extremely so, indeed, that it would amount even
to an impossibility. With respect to ventilation,
has it not been found, even when it has been attended to very particularly, to be, without the af-
fistance of other means, inadequate to the speedy
destruction of contagion? And, with regard to the
changing of wards, were it sufficient of itself to destroy contagion, might it not, I shall say sometimes,
from the number of patients received being equal,
nay even more than equal, to all the wards of
which the hospital consists, be utterly impractica-
ble? Other examples, and other queries, to the
same effect, were they not deemed superfluous,
might be advanced : then, considering the business
in this point of view, does it not become a duty in-
cumbent on us to look out for, and to try other
means more active, and more diffusive, which, with
the assistance of cleanliness, ventilation, changing
of wards, &c. may be employed for the purpose
of more speedily, and more effectually destroying
contagion? and may not the nitrous vapour of Dr.
J. C. Smyth, as I have already mentioned, be
deemed, of all other remedies extant, the most convenient, the most elegant, the most ingenious, and
the most efficacious for answering the wished-for
purpose, whether at sea or on shore?

The extraordinary effects which we have seen
the nitrous vapour produce, in cases of putrid ulcers, are facts of the utmost importance to mankind, and certainly deserve the most serious attention of medical practitioners. They not only shew,
in the most satisfactory manner, the power of that
vapour in such cases, but also point out, in my
humble

humble opinion, the probability of its having, in a similar way, as salutary a power in contagious fever, and in many other diseases proceeding from other species of contagion. This opinion may, perhaps, seem singular; but I shall endeavour to evince its consistence with reason and experience.

In hospital practice, it has been frequently observed, not only by me, but by other medical practitioners, that all the ulcers of patients in the same ward have on a sudden, and nearly at one and the same time, changed from, apparently, an healthy, to a foul, sloughing, or putrid state. I have bestowed considerable attention in observing this change; and, in the course of my practice, have been able to make the following remarks, which I shall here arrange as they stand among my memorandums.

That, first, one ulcer degenerated, then another, and so on, until all the ulcers, in the same ward had taken on a similar disposition.

That those ulcers nearest the one which first degenerated were sooner affected than those at a greater distance.

That this lamentable change did not happen in all the surgical wards at the same time.

That the patients, when their ulcers were in this degenerated state, laboured, more or less, under symptoms of fever, such as a frequent, small pulse, unnatural heat, sometimes chillinefs, dry skin, loss of appetite, &c.

That common dressings, common poultices, carrot poultices, turnip poultices, myrrh, Peruvian bark, applied to the ulcers, had no good effects.

That Peruvian bark, wine, opium, given internally, had, I thought, instead of good, bad effects.

P   That,

That, in one cafe, yeaft was tried, both internally and externally, but the difeafe evidently gained ground under the courfe.

That the acetum nitrofum,* whether ufed internally or externally, feemed to have good effects.

That the changing of the wards had always good effects.

That the nitrous vapour, with the like attention to cleanlinefs and ventilation as was in common beftowed, had, without changing the ward, as in the five cafes mentioned, as well as in many others, effects fuperior to thofe refulting from the changing of the ward, without the ufe of the nitrous vapour.

That the nitrous vapour had not the like good effects, without cleanlinefs and ventilation, as with them.

From thefe premifes, I have thought it warrantable to draw the two following conclufions:

1. That fuch a degeneration of ulcers, in hofpitals, from, apparently, an healthy, to a foul, floughing, putrid ftate, can only be accounted for on the principle of contagion.

2. That the nitrous vapour, with due attention to cleanlinefs, ventilation, changing of the wards, &c. is, feemingly, the remedy, of all others extant, beft calculated for preventing, or fpeedily deftroying that contagion; and from this naturally arifes the following query:

As, under fuch regulations, the nitrous vapour has fuch great power in preventing or deftroying one fpecies of contagion, may it not, under the fame regulations, be equally powerful in preventing or deftroying other fpecies of contagion?

I muft

---

* Vide Paterfon on Scurvy.

I muſt here obſerve, that the ſecond concluſion does not exclude the uſe of other medicines. Suitable remedies, both internally and externally, uſed at the ſame time with the nitrous vapour, will, no doubt, forward the cure. But, as theſe pages are intended for the purpoſe of pointing out ſome of the effects of Dr. J. C. Smyth's nitrous vapour, and not as a treatiſe on ulcers, I cannot, with reſpect to the latter, make, with any degree of propriety, an attempt on either the indications of cure or remedies.

On the preſent ſubject, I might, to what has been advanced, add many more medical obſervations, were I not of opinion that, after the experiments of Dr. Smyth,* they would appear ſuperfluous; and I might, with equal propriety, have recourſe to chemical reaſoning, were I not prepoſſeſſed with the idea that, conſidering what has been already ſaid, reſpecting Dr. Smyth's nitrous vapour, by that very ingenious chemiſt Mr. Keir, of Birmingham,† it would be extremely preſumptuous.

On the whole, and to conclude, I cannot help being of opinion, as well from the facts with which Dr. Smyth has favoured the public, and from what Mr. Keir has advanced, as from my own experience, that very great benefit muſt reſult to mankind from the *proper uſe* of the nitrous vapour, on board of ſhips, in hoſpitals, in priſons, in all places where people may be crowded together, and even in private families, in preventing and in putting a ſtop to contagion, as well as in mitigating and removing other diſeaſes, in which other medicines would not perhaps have the like good effects.
And,

---

* Vide Dr. J. C. Smyth's letter to the Right Honourable Earl Spencer, &c. &c. &c.
† Vide Dr. J. C. Smyth's letter to the Right Honourable Earl Spencer, &c. &c. &c. Appendix.

And, therefore, I moſt ſincerely wiſh that the plan of Dr. J. C. Smyth may be univerſally adopted; and that it may, for the good of our navy and army, for the honour of our country, and for the benefit of mankind, be practiſed by medical men, and others, without their conceiving any prejudice againſt either it or its ingenious Author.

I have the honour to be,

Gentlemen,

Your moſt obedient,

Very humble Servant,

DAVID PATERSON.

Montroſe,
12th *Auguſt,* 1797.

POSTSCRIPT.

Since finiſhing the preceding letter, I have had an opportunity of making further trial of the nitrous vapour, in a diſeaſe of a ſingular nature. The hooping cough, which has been prevailing here all this ſummer, made its appearance in my family laſt month, this being a contagious diſeaſe, and a change of air having been found uſeful in removing it, I ſuppoſed that the nitrous vapour might not only operate in counteracting the contagion, but alſo have effects ſimilar to the changing of ſituation; and hence,

hence, that it might, providing the lungs of my little patients could bear it, prove a convenient, an
elegant, and ufeful remedy, on the prefent occafion.
It was from thefe conjectures, and knowing that it
would have been extremely inconvenient for me to
have fent my children from home, that I ventured
to make trial of Dr. Smyth's fumigating plan; the
refult of which I fhall ftate, as briefly as poffibly,
by fketching the following cafes.

My third child, a girl of five years old, was feized with a flight inflammation of the throat, attended with hoarfenefs, about the 6th of laft month;
and about the 10th with a cough on which the inflammation and hoarfenefs went off, the cough for
fome days, feemed to be of a common kind, from
cold; but by the 15th it affumed the appearance of
hooping cough, accompanied with a flight degree of
fever. By this time my fecond child, a girl of fix
years, and alfo, my fourth or youngeft, a boy of
fifteen months, had begun to cough.*

On the 17th, the third child had frequent and
violent fits of coughing, with the hoop ftrongly
marked; and the fecond and fourth, though not fo
ill, evidently laboured under the difeafe. In the
evening of this day, I began the ufe of the nitrous
vapour, I fhut up my little patients, with a fervant,
in their bed-room fixteen feet by twelve, and fix
feet nine inches high, myfelf fuperintending the bufinefs. Inftead of a pipkin for holding the hot fand,
I ufed an iron pot, in which was placed two gallipots, containing the concentrated vitriolic acid and
nitre, according to the directions of Dr. Smyth. In
about five or fix minutes the room was filled with
vapour, and continued fo for an hour, without any
of

* My oldeft child had the hooping cough about three
years ago.

of the children coughing or shewing any signs of uneasiness.

18th, 19th, 20th. The fumigation was repeated every evening, and continued an hour, without coughing or any uneasiness occurring.

21st. Now, all my little patients seemed better, the fits of coughing recurring less frequently, and the mucus being more easily discharged than before.

14th September. From the 20th ult. to this time, the fumigation was repeated only six times, exactly in the same manner as before, without the children complaining of, or apparently feeling any disagreeable effect from it; and now, on account of the mildness of the disease, the cough not being troublesome even in the night, it was discontinued. —At this time the youngest child evidently laboured under symptoms of teething; the cough, however, did not appear in the least aggravated.—To the second and third child, during the course of the nitrous vapour, no other medicine was given; but to the youngest, who was frequently constipated, a weak solution of antimon. tart. which always operated downwards, and sometimes upwards, was occasionally administered.

23d. The second, and third child, continue well; and the youngest, though much distressed with teething, coughs but seldom, and very gentle.

Now after having stated these facts, whether or not the nitrous vapour, had any effect in counteracting the contagion or otherwise rendering the disease mild, and of short duration; or whether or not the disease would have naturally appeared mild, and have continued but a short time, without the interference of art, are points which I shall not take upon me to determine. Further trials are undoubtedly necessary for the purpose of forming a judgment.

ment. It muſt be confeſſed, however, that, even
the few trials which have already been made, ſerve,
in the mean time, a very uſeful purpoſe; they
clearly ſhew, that even young children labouring
under a diſeaſe, in which there is always, more or
leſs, a determination to the lungs, &c. are capable
of inſpiring the nitrous vapour without feeling from
it any diſagreeable effects : hence there cannot, I
preſume, be any objections to further trials of it
being made in the hooping cough. Alſo theſe facts
lead naturally to other important inferences, but as
they muſt be to you ſufficiently obvious, I avoid
making them.

I have the honour to be, Gentlemen,
Your obedient humble ſervant,
DAVID PATERSON.

Montroſe. 25ᵈ *Sept.* 1797.

EXTRACT

OF A

L E T T E R

From Mr. Abraham Baſſan, *Surgeon of his Majesty's
Hoſpital Ship* Union, *to the Commiſſioners of Sick
and Wounded Seamen, dated the* 22d *of* November,
1797.

" I USE the nitrous fumigation every day through
the Ship, and, as formerly, in the ulcers, from the
Sandwich, I found they ſpread and became foul
from local debility, being never apparently benefited
by the bark, wine or generous diet, and I uſed the
nitrous fumigation to the ulcers, after waſhing them
clean, with great ſucceſs."

. *Three*

Three Letters from Mr. JAMES M' GRIGOR, Surgeon to the 88th Regiment, at Jersey. The two first addressed to Dr. CARMICHAEL SMYTH, and the last to Dr. GARTHSHORE, of London.

Jersey, October 8, 1797.

SIR,

As Surgeon to the 88th regiment, I have for these last four years, been witness to the dreadful ravages of an infectious fever in different quarters of the world. In England, in the island of Jersey, on the continent of Europe, during a voyage to, and in different islands of the West Indies, this fever has been the scourge of the regiment to which I have the honour to belong, and after a trial of every mode of practice which I could learn, it proved extremely fatal.

On my return from the West Indies, having seen in Duncan's Annals of Medicine, an account of your work on fever, I determined to take the earliest opportunity of giving a trial to the mode which you recommended, of weakening and destroying contagion.

The 88th regiment, for nine months previous to their landing in Jersey, had been in the most healthy condition; they landed on the 6th of June last, and continued in the same healthy state till the middle of last July. On the 17th of July, the first case of a fever, which has since very generally prevailed, made its appearance. The person was seized with the worst symptoms of low typhus fever, and died on the 5th day. Having four years before, in this island, in the course of ten weeks, lost fifty men in the same

fever,

fever, I determined to give the fulleſt and faireſt trial to any thing recommended by you. I had every aſſiſtance from one of the ableſt commanding officers, Col. Bursford.

To diffuſe the contagious poiſon, I ordered the men, on the firſt appearance of fever, to be moved from the barracks, (which are in an unhealthy ſituation) to tents pitched at ſome diſtance, on a dry and healthy ſpot.

To deſtroy the virulence of the contagion, where it evidently exiſted, I made my two mates regularly fumigate the barrack rooms and hoſpital with nitrous vapour, in the manner you direct, the event will ſhew the ſucceſs, of perhaps the firſt trial of your excellent invention, in the army.

Of fifty-four caſes of this fever, which occurred from the 17th of July to the 24th of September, the firſt is the only one that I have loſt. The very remarkable effect of the nitrous fumigation, appeared from the great diminution of the number taken ill, after it was uſed. During the firſt week twenty-four caſes appeared, in the ſecond week, ten, in the third, ſeven caſes; and to this date,* the number of caſes continue to leſſen.

The effect of the nitrous fumigation, is evident, not only in the diminiſhed number of caſes, but alſo in their degree of virulence. The caſes that have of late appeared, have been gradually becoming milder, and are now what a late writer would call caſes of ſimple fever, having neither petechiæ nor any dangerous ſymptom.

As your work on fever has but lately fallen into my hands, I have been able, only in few inſtances, to follow the practice which you recommend; every thing indeed has been ſo completely effected, by following

Q

* The 8th of October.

lowing your manner of deftroying contagion, that, in general, little has been left for me, but to obviate debility.

If you fhould wifh to make any ufe of this communication, or fhould favour me with any thing further regarding your truly valuable difcovery, pleafe addrefs for me to the care of our agent, A. M' Donald, Efq. Pall-Mall court, London.

In a Memoir which I tranfmitted to the Army Medical Board lately, after an account of the fever which appeared in the 88th regiment, I gave an account of the fuccefs of your plan, and thought it my duty to recommend it to the attention of the Board.

I have the honour to be
Sir,
Your very obedient humble fervant,
JAMES M'GRIGOR,
Surgeon 88th *Regiment.*
Dr. Carmichael Smyth.

Jerfey, December 9, 1797.

Sir,

By a letter of yours of the 26th of October, with which I was favoured, I was happy to learn that my conduct had met your approbation.

In September laft, I fent to my friend Dr. Garthfhore, a copy of my Memoir to the Army Medical Board on the fever that lately prevailed in the 88th regiment. At his defire, I about three weeks ago, fent him an abridged account of two memoirs on this fever, which, I believe he intends to tranfmit to Dr. Duncan, for the fecond volume of the Annals of Medicine.

Medicine. Prefuming that it would be fatisfactory to you to fee them, I herewith inclofe you copies of my firft memoir, and of the abridged account fent to Dr. Garthfhore. I had nearly finifhed a fecond memoir, and hoped to be able to have concluded my *statement* of the fever, with an account of its extirpation ; but a few more cafes have of late occurred. The appearance of thefe laft cafes is however very naturally and eafily accounted for, and the treatment of them adds a ftill further teftimony to the efficacy of the nitrous fumigation, and of the treatment formerly purfued.

About the middle of October laft, from the almoft total difappearance of the fever, we relaxed confiderably in the fumigation, the only cafes then in the fever-hofpital, were convalefcents from fever. About this time, the different encampments breaking up, we were obliged to admit a good many pneumonic and fome dyfenteric cafes which crowded this fmall hofpital, (an old farm-houfe hired for an hofpital) and I was alarmed to fee the fever break out again with all the original fymptoms in fix or feven convalefcents. By attention to the different circumftances regarding fumigation, following the former treatment, and thinning the different rooms, matters are now brought to nearly the fame good ftate as before the re-appearance of the fever.

The re-appearance of the fever, has however, not been entirely without its advantages, it has been the means of fhewing me a fact, which efpecially in military practice, I conceive to be of the firft importance. It has pointed out to me the efficacy of the nitrous fumigation in deftroying dyfenteric contagion. I have often been witnefs to the rapidity with which the contagion of dyfentery flew through the wards of an hofpital, and how apt convalefcents from other

difeafes,

diseases, and in particular from fever, were to be
seized with this disease. Though thirteen cases of
dysentery were sent to the hospital, and some of them
with very severe symptoms, I know only of two instances where the disease was communicated in the
hospital, and with the exception of two chronic cases,
the cure in all, has been much speedier than I have
formerly seen it under the same treatment.

Confident of success, I wish much for opportunity
of trying the fumigation in other contagious diseases,
particularly small-pox. In two cases of cynanche,
attended with low fever, I used it, and both patients
(officers) are now well. I should be glad to hear if
you have extended the trial to other diseases.

Some other circumstances regarding our fever, I
think proper to mention to you.

So sanguine was I at one time, in my expectations
from the fumigation, that in some cases of the fever
which I set apart, I trusted solely to nature and in-
dustriously fumigating, but I soon saw that in these
cases I was rapidly losing ground. I next in conjunc-
tion with the fumigation, followed Dr. Cullen's plan
of treatment; this in every case protracted the cure,
and in several instances, I was obliged totally to aban-
don the plan. Being in possession of such powerful
means of destroying the contagion, as the nitrous
fumigation, I ventured to take the opportunity of
giving trial to, and comparing several of the modes
of practice recommended in fever; but a comparison
of cases as nearly equal as could be collected, gives
the most decided superiority to that recommended
by Dr. Robertson of Greenwich. But I think it
likewise proper to mention, that under the type which
fever lately assumed in this island, however proper
the immediate exhibition of the bark was here, that
it was by no means found so proper a remedy in other
situations.

fituations. In the Weft Indies, every trial given to the bark in the yellow fever during the paroxyfm, failed, and this not only with me, but in the hands of the phyficians who had the charge of the largeft hofpitals there.

I fhall never fufficiently regret my being unacquainted with your difcovery, while in the Weft Indies; and though doubtlefs it will for a time, meet with the fate of every other that has been made for the benefit of mankind; yet, if candid and liberal practitioners will but do their duty, and give it a trial, I am confident it muft foon carry conviction, and that you will derive that credit from it, which you fo juftly deferve.

I have for fome time been in the practice of keeping journals of my cafes; every cafe of this fever that has occurred fince its origin, has been regiftered either by myfelf or by one of my affiftants. The garrifon furgeon here, and his affiftants likewife, have witneffed the treatment: and the different facts mentioned in my memoir, are not unknown to feveral practitioners in the ifland. Any information which you may require from me at any time, regarding the trials of the nitrous acid, and the other means recommended by you, I fhall moft readily give.

I have the honour to be

With the greateft refpect,

Sir,

Your moft obedient humble fervant,

JAMES M'GRIGOR.

Dr. Carmichael Smyth.

St.

*St Owen's, Jersey, Nov. 1, 1797.

SIR,

As perhaps the confirmation or refutation of an *opinion* in medical science, especially of one that so nearly concerns mankind in general, as a mode of obviating contagion, is not less useful than a new theory, a new medicine, or a new mode of curing a disease; I shall lay before you some facts, which, I think, confirm the method of obviating and expelling febrile contagion, which has been recommended by Dr. C. Smyth.

On the 17th of July last, a contagious fever of the typhus form, appeared in the 88th regiment in this island. In different situations, this regiment had suffered severely for the last four years, from a fever of the same kind. In this island, three years ago, I had the misfortune to lose, in the course of ten weeks, forty† men from this fever. Soon after our arrival in Jersey, in last June, and but a short time before the first case of this fever appeared, Dr. Smyth's work on fever came into my hands: I, therefore, on the first appearance of the fever, gave the fullest trial to the use of the nitrous acid, and as the result shews, with the best success.

The account which I here give, is extracted from two official memoirs, which I transmitted to the Army Medical Board on the subject.

The fever had nothing remarkable in its appearance from typhus fever, as I have usually seen it occur,

* This letter has been already published in the second volume of the Annals of Medicine, and is only republished here from a wish to bring the whole of the evidence on this subject into one point of view.

† There seems to have been a mistake here, as in his first letter to me, he states the number to have been fifty.

occur, if I except the fuddennefs of the attack, often with delirium or epilepfy ; a very remarkable degree of debility ; and a great proneuefs to relapfe. Two cafes had the fcarlet eruption with angina, and the moft of the firft cafes, until they were fent to camp, and were expofed to a current of air, had petechiæ. In moft of the frft cafes, the contagion could be clearly traced.

On the firft appearance of this fever, alarmed at the fatal iffue of the firft cafe, I myfelf not only carefully fumigated the hofpital, the clothes, and bedding of the fick, with nitrous acid, but my two affiftants, Meffrs. Bruce and Brown, likewife conftantly fumigated the different barrack-rooms*.

Every man as foon as feized with the fever, was removed from the barracks, which are unhealthily fituated, to tents pitched in a dry and airy fituation, about a mile diftant. The barracks were likewife thinned by encamping or removing from them, near half the regiment.

The *treatment* in general was, by immediately exhibiting the bark after giving an emetic or cathartic, and afterwards giving cordials, blifters, &c. as indicated. The lavatio frigida, in feveral cafes, feemed to anfwer exceedingly well.

The *firft* cafe, which appeared on the 17th of July, died on the fifth day, with fymptoms which are ufually called highly putrid. This excited a very general alarm in the regiment.

I fhall here from my journal of cafes, and the copies of reports made to the Medical Board, give a ftatement of the appearance of the cafes of the fever : nothing can afford ftronger proof of the efficacy

---

* The rooms, by this procefs, were rendered fweet, and the men themfelves foon became fenfible of the comfort of it. Vide firft Memoir fent to the Army Medical Board.

efficacy of the means adopted to obviate and remove this difeafe.

From July 17 to July 28,	20	cafes appeared.
From July 29 to Aug. 4,	16	—— ——
From Aug. 5 to Aug. 11,	10	—— ——
From Aug. 12 to Aug. 18,	8	—— ——
From Aug. 19 to Aug. 25,	3	—— ——
From Aug. 26 to Sept. 1,	2	—— ——
From Sept. 2 to Sept. 8,	4	—— ——
From Sept. 9 to Sept. 15,	1	—— ——

From this time, till the fick were removed from
the tents, to a houfe where they were crowded,
hardly another cafe appeared. At this period,
however, about the 20th of October, the fever with
the original fymptoms again appeared among fome
convalefcents and pneumonic cafes; but by putting
in practice the nitrous fumigation, and thinning the
wards of the hofpital, the fever was again very foon
got under.

Of the total number of cafes of this fever that occurred, viz. fixty-fix,* the firft is the only one that
was loft; this, no doubt, is remarkable, and I not
only afcribe this fuccefs to the ufe of the oxygenated
nitrous acid, but I likewife think it highly probable,
that by fumigating with this acid, the contagion is
now nearly extinct, and that by its ufe, more cafes
of this fever have been prevented from appearing.

I have nothing at prefent to add to what Dr.
Smyth has faid of the ufe of the nitrous acid; every
trial which I have made hitherto, tends to confirm
his experiments. I have fet on foot a trial of different other acids,† which however is yet fo incom-
plete,

* A very large proportion this of the regiment, which
confifted only of four hundred men.
† We attempted, likewife, the extrication of the muriatic
acid gas, but with great inconvenience, and it is obvioufly
not fo proper as the nitrous. Vide firft Memoir.

plete, as not to allow me to fay any thing of their comparative merits. Juſtice to the author of ſo valuable a diſcovery, and a wiſh to make more public, what is ſo intereſting to mankind; has induced me to fay ſo much.*

I am, &c.

JAMES M‘GRIGOR.

Dr. Garthſhore, London.

———————————

For the following COMMUNICATIONS, *I am indebted to my Friend Dr.* JOHNSTON.

Queen Street, Portſea, December 23, 1798.

DEAR SIR,

WHEN I had the pleaſure of ſeeing you at Portſmouth, I had experienced the good effects of the nitrous fumigation in arreſting contagion, and would have communicated my obſervations to you on this ſubject ere now: but unfortunately the diſeaſe has been *re-introduced*, by receiving men ill of typhus fever, and dyſentery, from the Hillſborough tranſport, bound to New South Wales with convicts. I am, therefore, ſtill purſuing Dr. Smyth's method of fumigation, and am happy to ſay, that, I have, a ſecond time, experienced the moſt happy and beneficial effects from it. I will, as ſoon as poſſible, forward to you, through Mr. Palmer, the particulars of my ſucceſs, and one advantage, at leaſt, will ariſe from this unavoidable delay, viz. the pleaſure

R   of

* I am very confident, that in various ſituations the attaching to every regiment one large tent, and allowing the materials for the nitrous fumigation, would materially beneſit the ſervice. Vide firſt Memoir.

of feeing the fumigation fucceed a fecond time, under the very unfavourable circumftances of *bad weather*, and direct communication with the faid tranfport, from on board of which twenty-two fick convicts have been received, and placed under my care at Langftone harbour. I believe if we had not received the faid twenty-two men, that we fhould not at prefent have a fingle fick man among the convicts.

I am, &c.

(Signed)   SAMUEL HILL.

Dr. Johnfton.

Queen Street, Portfea, January 13, 1799.

Dear Sir,

Herewith I have the honour of forwarding to you fome remarks I have made on the effects, very happy effects, of the nitrous acid fumigation, in ftopping the progrefs of a contagious fever on board the hulks in Langftone harbour; and I am happy to convey to you my decided opinion in its favour; an opinion, not haftily formed, but founded on facts, which occurred on trying that excellent method of preventing, and leffening fome of the miferies of our fellow-creatures.

I will efteem it a particular favour, if you will perufe the inclofed letter, and forward it to Dr. Smyth.

I remain, dear Sir, &c.

(Signed)   SAMUEL HILL.

Dr. Johnston.

Portfea,

Portſea, January 13, 1799.

SIR,

HAVING with great pleaſure peruſed your publication on the fever, which prevailed among the
Spaniſh priſoners of war at Wincheſter, and alſo,
your account of the ſucceſs attending the nitrous
fumigation, in deſtroying contagion on board the
Union hoſpital ſhip, and on board a ſquadron of
Ruſſian ſhips of war at the Nore; I determined to
try its effects on board the hulks in Langſtone harbour, near this town, the firſt opportunity that
might offer.

A fever of a contagious nature made its appearance on board the ſaid hulks, in the month of July
1798, which ſoon became alarming, not from the
number of ſick only, but from the rapidity with
which it advanced to its laſt, or fatal ſtage. The
number of patients continued increaſing from the
6th of July, to the 29th of Auguſt; in the former
month ſixteen were received, in the latter ſixty-ſix.
Upon my repreſenting to Mr. Dyne, contractor for
the care of convicts on board the ſaid hulks, that I
thought great benefit would enſue, if the method of
fumigating recommended by you, was put in practice at Langſtone harbour, he with a liberal hand
ſupplied the concentrated vitriolic acid and nitre,
and humanely ordered that no expenſe might be
ſpared in attempting to ſtop the progreſs of the
fever.

Pipkins, &c. were procured here without loſs of
time, and we began fumigating on board the Sincerity hoſpital ſhip, on the 29th of Auguſt, 1798,
at which time it contained fifteen very ill of fever,
fifteen recovering from fever, (and three of other
complaints left in July) thirty-one had returned to
the

the prifon cured, and five had paid the debt of nature.

It was with extreme pleafure I obferved the effects of the vapour on many of the fever patients. I will ftate them on one, which, with little variation, will ferve for the reft.

Daniel Stowell, aged twenty-feven, was received into the hofpital-fhip on the 21ft of Auguft, where he had continued getting worfe till the commencement of the fumigation; he was then in the following ftate. A moft fevere and confufed pain in his head, with intenfe heat on the fkin, and infatiable thirft; tongue rough and extremely dry, appearing like a burnt cruft, and of a blackifh colour, inability to put it out of the mouth when defired; teeth and gums covered with the fame kind of fur as the tongue. Pulfe one hundred and thirty-one, fmall and weak. The vapour made him cough very much, and he requefted (his own words) *to be fmoked no more;* it was, however, repeated three times this day. Auguft 30, Pulfe one hundred and one, and ftronger, fome moifture on his tongue and gums, heat on the fkin greatly decreafed; head-ach much the fame; thirft not fo intenfe: fumigated three times this day. Thirty-firft, He was in every refpect better; continued the fumigation three times. September 1ft, Pulfe feventy-nine, and ftill ftronger than the two preceding days; other fymptoms much relieved: continued the fumigation. Second, Pulfe fixty-eight, ftrong and regular; his appetite permitted him to take more nourifhment than he had been able to do fince his illnefs: from this day to the ninth, he continued getting better, and I now confidered him in a ftate of convalefcence.

It is proper here to remark, that this man had an ulcer on his right leg, extending from the outer ancle

ancle acrofs the anterior part of the tibia to the gaf-
trocnemius mufcle, at which part it was four inches
broad and very deep, and difcharged a thick ill-
conditioned matter, which was very offenfive. Its
length was rather more than fix inches. After fome
days fumigating, I obferved a change for the better
in the appearance of the ulcer, and as I had tried a
variety of methods to heal it, for many months before he became ill of the fever, without producing
any good effect; I laid this change to the effects
of the vapour. I now directed it to the ulcer
itfelf, and continued its ufe till October the 30th,
when his ulcer was cicatrifed; and I verily believe,
to its efficacy is owing this poor man's cure. He is
gone to New South Wales in the Hillfborough,
which failed juft before Chriftmas laft; I faw him
the day before he embarked on board her, and the
cicatrix was very firm: the ulcer was the confequence of an old gun-fhot wound.

There is another cafe of ulcer which I think has
mended greatly fince I fumigated it; I am proceeding as in the former cafe, and I have great reafon
to believe it will ultimately prove fuccefsful. The
only dreffings ufed during the fumigation, or more
correctly fpeaking, after the commencement of the
fumigation, were dry lint and ung. refin. flav. over
which was laid a rag of linen, conftantly kept wet
with aqua lytharg. acetat. diluted with water.

Finding fuch beneficial effects from the fumigation
on board the hofpital fhip, in bringing the fever
fooner to a conclufion, by fhortening all its ftages, I
determined to apply it to the fource of the contagion;
and accordingly the prifon-hulks, la Fortune and
Ceres were fumigated every night from October the
15th to November the 20th, (except two nights,
the fervant who was fent for the acid to town (Portfea)

fea) having ftaid the firft night at a public houfe, and did not arrive in time to fumigate the next night,) and I had the pleafure of finding the fick reduced to eight; and feven days had elapfed, and not one patient had been fent to the hofpital-fhip. The fumigation was now difcontinued.

November the 21ft, eight men were received from the Hillfborough Botany Bay fhip, one of which number was in the laft ftage of a contagious fever, and two laboured under dyfentery. Several patients in a ftate of recovery caught the new contagion, and many attendants were taken ill with the diarrhœa and dyfentery; and as it was impoffible to prevent communication with the prifon-hulks, the prifoners again became fickly, and many died; fome of whom were not ill three days before that awful event took place; and one man in particular died delirious twelve hours after he was received into the hofpital-fhip.

The fumigation, which I confidered as the anchor of hope, was again reforted to November the 26th, and continued to the prefent time, January * 13th, 1799; and it is with fuperlative fatisfaction I add, that there has not been a patient received for the laft eighteen days, neither is there a fingle fever patient in the hofpital.

It is not my intention, neither is it neceffary, to trouble you with a detail of my method of treating this fever; what I have to fay relating only to the happy effects of the fumigation, and it is with peculiar pleafure I affert, that the progrefs of the fever

* Dr. Smyth will be pleafed to add three days more to the time which had elapfed on Sunday laft (the 13th) without having received any patient whatever from the hulks into our hofpital-fhip —Poftfcript of a Letter to Dr. Johnfton, dated the 16th of January.

ver has been *twice* completely arrefted by perfeverance in its ufe; it may, however, be neceffary and proper to fay a few words on its firft introduction, and on its fatal effects on many who were its victims; I will, therefore, relate the cafe of one patient, which will fhew inconteftably the nature of the difeafe.

The firft perfon taken ill, was John Smith, a convict, who had been fent from Newgate a fhort time before. He informed me, July 6th, 1798, the day he was received on board the hofpital-fhip, that he had been ill of a fever in Newgate, and had not recovered his ftrength when he was fent to the hulks, at Langftone harbour. The convicts having been remarkably healthy, previous to the reception of this man, and becoming very fickly immediately after his admiffion, I confidered the fever as introduced by him. He complained this day, (July 6) of head-ach; pains in his back, loins, and extremities; his fkin was very hot and thirft great; tongue covered with a yellowifh mucus; countenance of a yellow tinge, and great dejection of fpirits; pulfe ninety-fix, and weak. Seventh, All his fymptoms feemed aggravated, his pulfe one hundred and feventeen. Eighth, He was delirious, and required, at leaft, two attendants to keep him in his cradle; pulfe one hundred and forty. Ninth, He was covered with myriads of petechiæ, and became extremely offenfive: I was now unable to reckon his pulfe. Tenth and Eleventh, His petechiæ had run into each other fo as to form large blotches. Twelfth, There was a great difcharge from his nofe and ears, of a very dark colour and very thin. Thirteen and fourteen, The ftench from him was offenfive in the extreme. On the fifteenth, he died univerfally convulfed.

The

The annexed table will fhew the numbers taken ill before and after the commencement of the fumigation. The ninth and tenth of November were the days on which it was omitted, and it is remarkable, that although on the former day no patient had been fent to the hofpital, on the two fucceeding days feven were brought each day.

I have not, Sir, the honour of being known to you; but liberality and love of truth induce me to forward to you the above particulars, which I hope will be received as a token of that efteem and regard with which

I am, Sir,

Your moft obedient humble fervant,

SAMUEL HILL.

Dr. Carmichael Smyth.

N. B. There were 748 convicts on board the two hulks on the 6th of July, 1798, of thofe, 418 were received into the hofpital with the jail-fever, befides 24 from on board the Hillfborough. Total 442, of whom 71 died of the diftemper.

A Monthly

A Monthly and Daily RETURN of the CONVICTS attacked with the Jail-Fever on board the Hulks, and received into the *Sincerity* Hospital Ship in Langstone Harbour, from the 6th July to the 26th Dec. 1798. By S. HILL, Surgeon to the Hospital.

Months	Days	Numbr	Month	Days	Numbr	Month	Days	Numbr	Month	Days	Numbr
July	—	16	Oct.	16	3	Nov.	1	1	Dec.	1	3
Aug.	—	66		17	2		2	1		2	2
Sept.	—	120		18	1		3	--		3	3
Oct.	1	5		19	2		4	1		4	5
	2	3		20	3		5	--		5	--
	3	7		21	4		6	3		6	5
	4	11		22	3		7	1		7	4
	5	3		23	1		8	1		8	3
	6	7		24	3		9	--		9	2
	7	3		25	2		10	7		10	2
	8	4		26	1		11	7		11	3
	9	2		27	2		12	3		12	2
	10	2		28	1		13	1		13	1
	11	6		29	1		14	--		14	2
	12	8		30	0		15	--		15	1
	13	3		31	1		16	--		16	3
	14	5					17	--		17	2
	*15	9					18	--		18	2
							19	--		19H	6
							*20	--		20	2
							21H	--		21	2
							22	--		22	3
							23	1		23	2
							24	7		24	1
							25	4		25	1
							*26	2		26	2
							27	3			
							28	2			
							29	2			
							30H	3			
To the 16th		77	To 31st		30			52			64

* On the evening of the 15th of October we began to fumigate the hulks, and continued to do so every day to the 20th of November, (the 9th and 10th of this month excepted;) on the 20th, the fumigation was discontinued, but resumed on the 26th, and continued without interruption to the 13th of Jan. 1799, though on the 26th of December the sickness had entirely ceased.

[H] On the 21st of November, eight persons ill with the jail-fever, or dysentery, were received into the hospital, from on board the Hillsborough Botany-bay ship, outward-bound; eleven more were received on the 30th of the same month, and five on the 19th of December: in all, twenty-four.

S

SIR,

Forton Hofpital, Jan. 17, 1799.

SIR,

A friend of mine having informed me that you were preparing for the prefs fome additional experiments to prove the utility of the nitrous acid fumigation, in deftroying contagion, I take the liberty of fubmitting to your notice fome facts which have occurred to me on this fubject. If you will honour them with a perufal, and think them of fufficient importance, I fhall feel myfelf highly flattered, fhould they, through your fanction, become more publicly known.

From the 4th to the 12th of April, 1797, we were employed at this hofpital in receiving prifoners infected with the jail-fever. Thefe men had recently arrived in fome tranfports from Wales: They formed part of a new regiment who had landed there, and who, previous to their failing from Breft, had been releafed from various prifons in France.

This fever was generally attended with petechiæ, and, in fome cafes, the parotids were affected. From thefe fymptoms, and from its being highly contagious, it was thought immediately neceffary to fumigate with the nitrous acid in greater quantity than we had hitherto done, and, inftead of *three* fumigating pipkins in each ward, we increafed them to *fix*. This method appeared to have a very defirable effect, for by filling a fever-ward with the vapour to fuch a degree that the fmalleft part of it could not efcape the influence of the nitrous gas, we never failed to render the air of that ward fweet and refrefhing for fome hours. Here I muft beg leave to remark, notwithftanding the prejudices that are abroad, that I never remember the vapour having appeared too powerful for any fever-patient to bear, and although phthifically inclined myfelf I could al-

ways

ways bear it without inconvenience. If, however, some inconvenience should occur, I conceive it only a secondary object of consideration, when compared with the good effects to be derived from the fumigation, in checking contagion.

The increased demand for the nitrous vapour, from the number of wards occupied in the hospital, occasioned the consumption of nearly three pints of the concentrated vitriolic acid, and a proportionate quantity of the purified nitre, each day. The expense of so increased a consumption, if an object, is certainly over-balanced by the considerable advantages resulting from its use; for, notwithstanding fresh contagion was every day brought into the hospital till the month of July, by means of prisoners received from the prison-ships, we had the satisfaction to see the malignity of the fever subdued; and by the unremitted attention of the physicians of the royal hospital at Haslar, with a due observance of cleanliness, we were authorised early in the month of August, to report to the commissioners for sick and wounded seamen, that the jail fever which lately raged among the French prisoners at this port, existed no longer. The prisoners then, in proportion to their number, enjoyed an usual share of health, and which, previous to the introduction of the jail-fever, had been equal to the state of health of any number of working men in the manufacturing towns of England.

During the continuance of the jail-fever, *very few* of the establishment of the hospital caught the distemper, considering the length of time it prevailed at Forton. Those who received the contagion first, and were fond of drinking, either died, or had a very severe illness. Others, not so partial to liquor, and

and who received the infection later, in general recovered.

No putrid disease of any consequence made its appearance again in the hospital until the 3d of January, 1798, when we received some prisoners from his Majesty's prison-ships, Fame and Portland. These men, who had lately arrived in a transport from Falmouth, were affected with jail-fever; and on many of them petechiæ appeared. I immediately acquainted the commissioners for sick and wounded seamen of the circumstance, who gave orders to the physicians of Haslar hospital to visit Forton, and report to them accordingly.

At this time the government of France had agreed to victual, clothe, and attend in sickness, their own prisoners. The sub-contractor in the medical department did not, however, take charge of the hospital at Forton, till the 11th of February following. We had, previous to that time, an opportunity of fumigating with the nitrous acid with apparent good effect. The business of the hospital from that time, came under the guidance of the French, but although they had two hundred and eighty-two patients put under their care, and upwards of one hundred of them affected with putrid fever, the acid vapour was totally discontinued for near five months. The state of the hospital, from the 11th of February, was highly alarming, and extremely unfavourable to the health of the prisoners; the number on the hospital books increased daily, the wards were so neglected that they soon appeared dirty, and had an offensive smell; the œconomy of the buildings was not at all studied for the advantage of the sick; fever and convalescent patients were not separated, and in the course of twenty weeks two hundred and thirty-five prisoners died,

five

five hundred and thirty-feven remained fick in the hofpital; and the jail-fever was raging with great violence.

On the 20th of May, Dr. Forzy, and Mr. Brunet, Surgeon, arrived here from France, who were appointed by the French government, infpe
ors over their fick and wounded prifoners in England. Thefe gentlemen were placed at this depôt to fee that juftice was done to the French prifoners. As they were verfed in the management of hofpitals, they foon perceived the great inattention that had been paid to the fick; and the irregular manner in which the duty of the hofpital was carried on. They foon convinced Mr. Vochez of this fituation of the fick prifoners, who loft no time in breaking the contra
, and putting the management of the hofpital into other hands, thinking it preferable to fubje
 himfelf to the penalties of the law, than that one helplefs prifoner fhould fuffer by the hand of infatiable avidity and unfeeling negle
; a tranfa
ion highly honourable to Mr. Vochez, and the memory of which fhould always remain ftrongly impreffed on the minds of thofe who were, or might be benefited by it.

On the 1ft day of July, Dr. Forzy and Mr. Brunet commenced the management of this hofpital in the medical department. Their chief obje
 was to put the buildings into a general ftate of cleanlinefs, and to fumigate the wards of the hofpital with the nitrous acid. For this purpofe Mr. Vochez very liberally purchafed all that remained in the Englifh ftore, and likewife fent down from London a large quantity of the concentrated vitriolic acid and purified nitre. The acid vapour, together with the new fyftem of cleanlinefs, at the expiration of even fo fhort a period as feven days, caufed a very fenfible change

change in the appearance of the patients, and the deaths, which during the previous week amounted to *eighteen*, were reduced to *six* the firſt week after the new adminiſtration took place. Every ſucceeding week a more healthy aſpect preſented itſelf. At the end of the tenth, the hoſpital might be perceived to have nearly approached a ſtate of general convaleſcence. The number of contagious patients received from Portcheſter Caſtle, and the priſon-ſhips, were indeed conſiderably leſſened; yet the deaths which had occurred, were *ſixty-nine* in number, although the patients in the hoſpital were reduced to two hundred and ſeventy-three. In the next enſuing ten weeks, the deaths decreaſed to thirty-ſeven; the number of patients was two hundred and eighty-ſix. At this time all contagion had left us, and we had not received a patient with putrid fever, from Portcheſter or the priſon-ſhips, for near three weeks; and I have great pleaſure in ſtating, that we have remained in the ſame healthy ſtate ever ſince, and have not more than three hundred ſick in the hoſpitals of Forton and Portcheſter, though they contain the ſick of near ten thouſand priſoners at this port.

I cannot help expreſſing the great praiſe that is due to Dr. Forzy and Mr. Brunet, for their very meritorious exertions in forming ſuch ſalutary arrangements, and putting the hoſpital into ſuch excellent diſcipline and order, and particularly for their uncommon zeal, and the indefatigable pains they took to ſee the hoſpital well fumigated with the nitrous acid, which there cannot be a doubt was exceedingly inſtrumental in crowning their urgent endeavours with the deſired ſucceſs.

From the foregoing ſtatement, I think no one can heſitate to beſtow the high encomiums that are

due

due on this moſt excellent invention and its ingenious author, to whom every member of ſociety ſhould always think himſelf highly indebted.

I am, Sir,

With great reſpeſt,

Your much obliged humble ſervant,

JOHN   GRIFFIN.

Jas. Carmichael Smyth.

———————

Portſea, December 16, 1798.

DEAR SIR, .

IN anſwer to your favour of the 11th inſt. I have very little to add to what I have ſaid in my weekly reports, wherein the obvious difference of thirty the firſt week, and only three the next, being attacked with typhus, I in a great meaſure attribute to the nitrous fumigation, uſed in the way recommended by Dr. J. C. Smyth, from which it alſo appears, that of thirteen ill in this latter time, I had only oc-caſion to ſend two to the hoſpital, and that the infec-tion was ſo far ſubdued, that even nine got well on board, by the emetics, antimonials, &c. uſed in the firſt ſtage of the diſeaſe, which I had not been able to accompliſh previous to the uſe of the nitrous fumigation ; I muſt alſo add, that the keeping fires conſtantly burning, and the ſending thoſe to the hoſ-pital who were firſt taken ill, might have ſome good effeſt ; but the weather was ſo inceſſantly bad, we could not take advantage of wind ſails, and indeed laboured under every other inconvenience ; the De-fiance's ſhips company at this time, being on board the Elephant (a ſeventy-four) as a hulk, in the harbour.

bour. I imagine that the typhus fever was introduced into the ship by women, two of whom I sent on shore as soon as I discovered, one with a scarlet eruption, and certain degree of fœtor—Both Dr. Lind and Dr. Hope, are of opinion, the fever was of a very dangerous and infectious nature, and from Dr. Hope's opinion, I was more particularly led to be very attentive in removing the sick early to the hospital, but the nitrous vapour was scarcely used two days, when the symptoms abated, and has now entirely ceased.

The method I used, was by holding the pipkins under the hammocks of those with feverish symptoms, and at night, eight were carried about the decks, when all the hammocks and people were below, and this was attended with very little inconvenience to those in health. I have also used the nitrous vapour in the manner mentioned by Mr. Paterson for ulcers and foul sores, and I think, with obvious good effect, at least the patients themselves acknowledged it. I in general have a pot in fumigation, when I visit the sick in the cock-pit, which I really think, independent of the utility of applying sores over it, tends to purify the air and dispel fœtor, its smoke is particularly pleasant to me, and I often have it in my cabin, when the sick are below. I shall be happy if these cursory observations appear to you at all satisfactory.

And I ever am,

Your obedient servant and friend,

JAMES GLEGG.

J. Johnston, Esq.

Norman

Norman Crofs, Auguft 8, 1798.

Dear Sir,

AGREEABLY to the wifh you exprefſed, of being informed of any obfervations I might happen to make on the effects of the nitrous acid fumigation, in checking or deftroying contagion ; I have for a confiderable time paft, carefully attended to the confideration of that fubject, efpecially fince I read Mr. Paterfon's ingenious letter addreffed to the Commiffioners of the Sick and Wounded Board : and although I have not been able to draw fimilar concluſions, with refpect to its beneficial effects on patients afflicted with ulcers, it is becaufe I was lefs obfervant and attentive perhaps, to that point, not having perufed this gentleman's publication previous to my adopting a more fuccefsful mode of treating ulcers, I hope than is generally practifed or known.

Whether the nitrous vapour has peculiar fpecific powers for deftroying contagion, is difficult to fay, I have conftantly fumigated the hofpital wards three times a week, fometimes oftener, and I have ftrong reafon to believe that the nitrous acid fumigation not only tends to prevent the fpreading of contagion ; but anfwers other falutary purpofes.

Among the various trials I had occafion to make with the intent of afcertaining its effects on putrid effluvia, a circumftance occurred worthy of notice. The water clofet adjoining the agent's office, became in the very hot weather of June and July laft, fo extremely offenfive, that Mr. Perrot and his clerks complained, that unlefs the fætor was in fome degree removed or mitigated, it would be impoffible to continue much longer in the apartment. It is a room of twelve feet fquare : I ordered three pipkins in, and proceeded to fumigate for an hour, filling the place completely with the nitrous vapour. During

T this

this period, the door was unavoidably opened feveral times; but the agent, two clerks, and myfelf, continued there the whole time. It at firft caufed a little coughing; but in a few minutes that irritation ceafed, and the vapour became rather grateful than otherwife. At the end of forty-five minutes, Mr. Gardner one of the clerks, declared he had felt for fome time, an unufual fenfation of hunger, which at the conclufion was extremely importunate, and he that day ate, as he expreffed himfelf, voracioufly. Mr. Richards the other clerk, felt nearly fimilar effects, though not fo much expofed to the vapour as the former, who kept ftirring one of the pipkins frequently. As I did not feel fenfations of a fimilar nature at the time, I paid little attention to their obfervations; but on returning home, fomewhat better than an hour after, my ftomach became equally importunate and craving for food, in fo much fo, that it was with the utmoft difficulty I could refrain till the ufual hour of dining; and I may venture to fay, that I ate nearly twice my ordinary quantity on that day.

Ten days after this I had the wards more than ufually fumigated, in which I remained till the expiration of the procefs, and I felt equally affected with hunger, but although a more than common degree of flatulency attended it. I ought to mention that the agent's apartment was rendered perfectly fweet and pleafant, the fœtor being wholly removed or deftroyed. Mr. Perrot was fo fenfible of the change, that he requefted fome days after, to have the fumigation repeated, becaufe the fmell had again become offenfive.

Do not thefe facts evince the propriety of diffufing the nitrous vapour copioufly in the convalefcent wards, where lofs of appetite, and want of tone in the

the digeſtive organs, often retard the perfect reſtora-
tion of health?

Another circumſtance well deſerving our regard,
with refpect to the effect of the nitrous fumigation on
contagion, occurred. Some time ago we had nearly
forty patients in the Dutch hoſpital dangerouſly ill
with typhus fever; old men who had been recently
captured in the Greenland ſhips, and who laboured
under great dejection of ſpirits. They were taken
ill in the priſons, ſhortly after their arrival at Nor-
man-Crofs, and although we had a great many patients with ſlight complaints in the hoſpital, on the
admiſſion of thoſe with typhus fever, yet not one of
them, nor of the nurſes, caught the diſeaſe.

This was not the caſe in the priſons, for ſeveral of
the old ſtandards there, were attacked with the fever. During the continuance of the diſeaſe, we were
unremitting in our attention to fumigate the wards
of the hoſpital daily, with the addition of ſeveral
pipkins more than were commonly employed.

Are we not warranted then, in concluding that
theſe prophylactic meaſures, prevented the contagion
from ſpreading and infecting the patients in the hoſpital; whilſt thoſe in the priſons not having equal
advantages, were attacked with the fever?

We were however fortunate enough to loſe no
more than one patient, who died on the ſeventh day
in a highly putrid ſtate; a clear proof of the malignant tendency, and contagious nature of the diſor-
der.

I could enumerate a variety of other inſtances,
where it appeared to me that the nitrous acid fumigation was attended with ſalutary effects; but what
has been already ſaid is, I hope, ſufficient to ſhew
that many benefits are likely to accrue from a more
extenſive uſe of this medicine; and that it ought
almoſt

almost on every occasion, to have the preference to all other modes of fumigating, where contagion is prevalent.

It might certainly be used with much propriety and advantage in the sick births, and between decks, on board his Majesty's ships of war. It is true, where great moisture prevails, which is unavoidably the case in men of war, fires are absolutely necessary; after which the nitrous gas should be liberally diffused throughout the ship: but prejudice, which frequently warps the best understanding, prevents many from adopting salutary measures, because these means had not been originally suggested by themselves, or because they happen to militate against some preconceived and favourite opinion.

But before gentlemen decide on any remedy held out to the public in a fair and candid manner, they should first subject this remedy to actual experiment, and then determine for or against it, according to the unprejudiced result.

I do not pretend to say that the nitrous acid ought to supersede all other preventive remedies; on the contrary, ventilation and cleanliness must likewise be conjoined with it: but where contagious fever prevails, there is every reason to believe that these will prove the most effectual external means to subdue it.

I thought it my duty to say thus much in justice to the ingenious inventor of the nitrous acid fumigation.

I am, with the greatest respect,
                    Dear Sir,
            Your very humble and
                    Obedient servant,
(Signed)   JAMES MAGENNIS.

Dr. *Johnston*.

                                Extract

Extract of a Letter from JOHN SNIPE, *Efq. formerly Surgeon of the Sandwich, now Surgeon to the Naval Hospital at Yarmouth. Dated Yarmouth, June 17, 1798.*

ON the 9th of March, when I joined the Sandwich, a contagious fever raged in that ship, there were daily ten, twelve, or fifteen men fent to the hofpital with fever and ulcerated legs. You are already acquainted with the fteps that were taken to fubdue this. Our fuccefs I attribute to cleanlinefs, free ventilation, and the diffufing daily the nitrous gas through every part of the fhip, and I am convinced, that had that fhip been fumigated three or four times a week, with the nitrous vapour, no fuch fever would ever have been generated, notwithftanding the great number of men that were on board. Suffice it to fay, that in three months, this fhip was as healthy as any other of her clafs, although we never had lefs than 1000 and often 1500 men on board. During the laft twelve weeks that fhe was in commiffion, I only fent eight men to the hofpital. When I firft joined her, the fmalleft fcratch rapidly degenerated into a foul ulcer, but after the febrile contagion was fubdued, fores healed as kindly as in any other fhip.

About three weeks before fhe was paid off, an Italian cut one of the feamen in the thigh with his knife, the wound was about fix inches long, and nearly two deep, it was ftitched up, and a doubleheaded roller applied; it healed by the firft intention, which was a proof that no contagion remained in the fhip.

From fome experiments I have made, I am induced to think that febrile contagion, and the general exhalations from the human body, are of an

alkaline

alkaline nature, and that the nitrous vapour equally
fubtile, penetrating into every crevice or corner,
wherever it meets the former, deftroys its peftiferous
qualities, and renders it as inactive as a drop of
water. I have repeatedly condenfed the breath and
perfpiration of patients in typhus fever, and upon
adding an acid to it, an effervefcence was vifible:
the method I took to obtain a fufficient quantity of
liquid for the experiment, was to make the patients
breathe on cold panes of glafs, and to put the fame
under the bed-clothes, clofe to the fkin, when in a
ftate of perfpiration; in this way half an ounce may
be procured in a fhort time.

Fumigating with the nitrous vapour cannot be
too ftrongly recommended on board of all fhips, in
barracks, hofpitals, prifons, clofe cellars, and houfes
that are not conftantly inhabited. Some time ago
I had occafion to go to a ftore room, that had not
been looked into for three weeks or a month, it
contained the bedding ufed by the Dutch prifoners
before they were fent to Holland, the whole of which
had been fumigated, expofed to the fun for two days,
afterwards wafhed and perfectly dried before they
were put in ftore; yet the room fmelled very badly.
I immediately ordered it to be fumigated with fix
pipkins, and next morning it fmelled perfectly fweet:
for which reafon I have, and mean to fumigate all
the ftore-rooms, every week.

Laft winter I tried the nitrous vapour with fome
of the worft ulcers that perhaps were ever feen in
any country; the pipkins were placed under the
naked fores. It gave fome pain, and I could not obferve any good effects from it, but I beg leave to obferve that this was not the fault of the remedy, but
the untractable nature of the difeafe, for in many
cafes, no internal medicine, nor external application,
had

had the leaſt effect, amputation alone ſaved the patient's life. I mentioned this in a letter to Dr. Blane,
at the time, but ſince then, I have found it of eſſen-
tial ſervice in cleaning foul ulcers, buboes, and
ſtumps; and I do not heſitate to give it as my opinion, that if it is properly applied in time, it will
be found to be of more ſervice than moſt of the external applications that are at preſent uſed.   You
are well aware that in many caſes, we have to lament
the failure of every exertion.

It is with heartfelt pleaſure I ſay, that I have loſt
very few in fevers, ſince I have been at Yarmouth,
although we have received a great number in the low
typhus fever into the hoſpital; nor is there one inſtance of the contagion being communicated to the
ſurrounding patients, which I attribute to the nitrous
gas, cleanlineſs and free ventilation.

When the wards were filled with the nitrous vapour, patients who had weak lungs, coughed vio-
lently at firſt, but they breathed more freely afterwards, eſpecially if the atmoſphere was thick and
heavy, but I never obſerved any bad effects from it,
even with patients in the laſt ſtage of conſumption.
The nitrous vapour has a moſt aſtoniſhing effect in
correcting the factor in the wards, ariſing from extenſive bad ulcers : this alone is a great comfort both
to the patients and attendants.   I have ſo high an
opinion of the nitrous vapour for deſtroying contagion, and as a corrector of foul air, that I have in
the ſtrongeſt manner recommended the uſe of it to
ſeveral of my friends, in the different factories in the
Mediterranean.

There are many navy ſurgeons prejudiced againſt
the nitrous fumigation, beſides it gives ſome trouble, which will at all times operate powerfully with
the indolent.

Much

Much praise is due to the ingenious author of the nitrous vapour. I have not the honour of his acquaintance, nor did I ever read a word he wrote (for which I blush) on the subject, but I should be wanting in candour if I did not faithfully relate facts as I found them; and in my opinion, the nitrous vapour tends greatly to destroy contagion, and is a most powerful corrector of foul air. Query, Whether or not does it give an additional quantity of oxygen (pabulum vitæ) to the surrounding atmosphere, where it is diffused.

I do not expect to have fewer patients in the hospital than we have at present: there are fifty-three ships of war employed in the North Sea, and they never go to Sheerness but when in want of repairs. When the fleet comes in, I expect to get fifty or sixty fresh patients, and they are weekly sending in sick by the frigates and cutters. There are only two men in the hospital that were wounded in Lord Duncan's action, and the bad ulcers that were received last winter are mostly gone, yet the number is still kept up. There are more cases of ulcer received into this hospital, in proportion, than into the two royal hospitals, ulcers as well as pectoral complaints being more frequent in the North Sea, than in the Channel. I have also remarked that those who had the most obstinate ulcers had been for some years during the war in the East or West Indies, or in the Mediterranean.

I have an ample field here for observation, but little time to put my observations on paper, as between the hospital, sick quarters, and prison, I am kept constantly employed.

(Signed) - JOHN SNIPE.

Dr. *Johnston.*

Extract

Extract of a Letter from I. Blatherwick, Esq. Surgeon. Dated Farham, June 17, 1798.

SIR,

NOT having taken minutes of any observations I may have forwarded to you on the subject of the nitrous vapour, I am not enabled to state with the precision I could wish, the particulars relating thereto.—At the time we began to use the nitrous vapour as a fumigation, our hospitals were free from any contagious disease, and on the whole, as healthy as I ever knew them. But we never had so long a continuance of that healthy state, as while this vapour was used in the hospital; and it deserves notice, that soon after it was discontinued, under the French administration, the typhus fever again made its appearance with considerable severity. No instance, to the best of my recollection, has occurred of that disease being communicated either to nurse or patient, whilst the nitrous fumigation was employed; whereas many instances occurred from time to time, previous to its use. The difference on entering a surgery ward, after using the fumigation, is more remarkable than any other, as it deprives those wards of the smell peculiar to them. I have also remarked, that fewer patients, with extensive ulcers, whilst this was used, became hectic, than formerly. In regard to consumptive patients, I am not at present qualified to speak decidedly from any proofs I recollect of its effects, but judging from analogy, I am induced to believe that patients of this description are as likely to receive relief (in so far at least as the nature of the complaint admits) as patients in any other disease; for whatever dislodges from the air, or corrects the noxious particles accumulated from the effluvia of diseased bodies, must mitigate the symptoms of this

U   disease.

difeafe. No arguments I am mafter of can induce the French furgeons to adopt it ; they complain of its exciting cough, and injuring the catarrhal complaints. They content themfelves with burning a few pounds of juniper berries, daily in each hofpital, notwithftanding the effect of the difcontinuance of the former practice, has been evident enough to convince any unprejudiced perfon which deferves the preference. On the whole, I infer, that the *nitrous vapour* is poffeffed of ftrong antifeptic qualities—is capable of being adminiftered in every difeafe—can be procured with facility in all places—requires no extra information in the operator ; and, in fhort, is the beft fumigation I am acquainted with, to be employed where the patients remain in the wards.

(Signed)   I. BLATHERWICK,
                          Superintendant, &c.

James Johnſton, Eſq.

Letter from Captain Lane, of the Navy. Dated Plymouth, June 19, 1798.

SIR,

HAVING been particularly engaged with the Admiral for fome days paft, I have not been able to anfwer your letter before, otherwife I fhould not have been fo tardy in expreffing the fatisfaction I feel, in offering my teftimony of the apparent advantages derived from fumigating with the nitrous acid, in checking at leaft, if not in ftopping contagion. Had I fuppofed that the fmalleft doubt remained on the fubject,

fubject, I fhould have kept a very minute account
of every experiment, but as I conceived the thing
to have been uncontrovertibly proved before, I fa-
tisfied myfelf with having recourfe to it, whenever
there was any appearance of occafion for fo doing,
and I can with fafety fay, without ever having been
difappointed. On receipt of your letter, I immediately fent it to Mr. Harris, who has returned it to
me, and who I find is exactly in the fame predica-
ment as myfelf, having been prevented making any
particular remarks, by entertaining the fame opinion
of its efficacy as I had, and having been equally fa-
tisfied with the refult of his experiments.

With regard to the comparative degree of ficknefs
between the prifons and fhips, I muft refer you to
the weekly returns. I can however fay, that when
I firft obtained Dr. Smyth's fumigating materials for
the latter, a typhus fever was raging on board the
Prudent, which in Mr. Harris's opinion, was fo
far alarming as to induce me to remove the people
from her for a few days, during which time I had her
twice fumigated and the decks white-wafhed, after
which, the fame prifoners were returned to her, and
fhe has fince been as healthy as the other fhips.
When it is confidered how hot the weather has been,
and that I have at times been obliged to put as many as 600 prifoners in the fixty-fours, I am induced
to attribute the prefervation of their health, to the
effect of the fumigation, which regularly takes place
every Thurfday morning in all the fhips when the
weather permits, having long fince eftablifhed a fig-
nal to enfure its never being neglected in any of them.
With refpect to the prifon and hofpital, I am informed by Mr. Harris that whilft he had the materials
for the nitrous fumigation, it was made ufe of when
wanted, and always with the defired effect, but on
the

the French government taking charge of the sick, the fumigating materials were returned to the Royal hospitals along with the other stores.

Sorry I am indeed not to have it in my power to be more particular, for nothing can be more satisfactorily proved in my mind, than the efficacy of Dr. Smyth's means of checking, if not of destroying contagion, and I cannot but congratulate that gentleman on a discovery which I certainly consider as highly beneficial to mankind, as it must be grateful to his feelings.

(Signed)   CHA. HEN. LANE.

James Johnston, Esq.

Extract of a Letter from Alexander Brown, Esq. Surgeon of the Royal Sovereign. Dated Torbay, May 27, 1798.

As a fumigation for a sick room, I consider the *nitrous vapour* as an elegant, ingenious, and useful one. In the morning whilst dressing the sores, two pots are employed for fumigating them, the smoke or vaporous gas evidently sweetens the air, and I do not observe that any particular irritation of the lungs is excited; I have two patients labouring under phthysis pulmonalis, they are commonly placed near to the persons fumigating. I do not find that they cough more than usual, unless they hold their heads over the fumigating vessel. I commonly fumigate the sick birth every night, and the people who sleep there, say the place feels wholesomer than

it

it does when not fumigated.   As for its producing contagion, I have no reason to believe it does, at least I have never as yet found correctors or sweeteners of foul air, have that effect.

Some of my brethren here declare that there never was such a heavenly discovery for the cure of ulcers of all kinds, as the fumigation with the nitrous gas ; and that it possesses a singular healing power (in all habits) which never fails ; if so, I am singularly unfortunate.   I have heard of Paterson's pamphlet sent to the Captains and Admirals ; I am promised a reading of it ; I shall then be able to judge how far I have administered the vapour with propriety.

(Signed)   ALEX. BROWN.

Dr. James Johnston.

Extract of a Letter from John Drew, *Esq. Surgeon,* dated London, *the* 17th *of* June 1798.

Allow me to lay before you the following observation on the salutary effects of the nitro-vitriolic fumes, of which I was an eye-witness, whilst Surgeon of his Majesty's ship l'Unité.   The ship was in general unhealthy, but there were two men in particular ill of putrid fevers, who lay on the lower deck ; by constantly fumigating under their hammocks, I found the disease so much checked in its progress as not to extend to any other part of the ship's company.

(Signed)   JOHN DREW.

Dr. Johnston.

Extract

[159]

EXTRACTS

FROM THE

LETTERS OR JOURNALS

OF

SURGEONS OF THE NAVY,

(On the Subject of the Nitrous Fumigation;)

TRANSMITTED TO

DR. CARMICHAEL SMYTH,

BY ORDER OF THE

BOARD FOR SICK AND WOUNDED SAILORS.

Extract from the Weekly Return of Mr. George M'Grath, *Surgeon of his Majesty's Ship Ruffel, dated Oct.* 26, 1796.

Since we have had the Dutch prisoners on board I have found particular benefit from fumigating with the nitrous acid in purifying the foul air, and preventing contagious fevers, which otherwise would have originated from the uncleanliness, and filthy indolent dispositions of those men.

Extract

Extract from the Journal of Mr. John Drew, *Surgeon of his Majesty Ship,* l'Unité, *between the 22d of November,* 1796, *and the 28th of June,* 1797.

All the intermittent fevers and aguiſh complaints moſtly proceeded from the badneſs of the weather, and the lowneſs of, and dampneſs of the Unité, as the breathing of the people, and the wetneſs of the ſhip produced an infectious kind of air, which never failed to cauſe numbers to be taken ill, notwithſtanding the uſe of ſtoves, and conſtant fumigation compoſed of vitriolic acid and nitre, which I found to be very uſeful, particularly in two caſes of putrid fevers, as by conſtantly fumigating the place where they lay, the infection did not attack any more of the ſhip's company; the exhalations, I think would anſwer better if the fumigating pots were of another conſtruction, which I refer to your determination.

Extract from the Journal of Mr. James Runcie, *Surgeon of his Majesty's Sloop,* l'Eſpiegle, *between the* 19th *of February,* 1797, *and the* 10th *of February,* 1798.

We had a great many ſcorbutic caſes but they were ſo ſimilar in ſymptoms and treatment that I have not thought it worth while to mention them; there were alſo a variety of febrile and other complaints, of which the above are the moſt conſiderable: the fever introduced by the French priſoners we had a great deal of difficulty in ſubduing, as the infection ſpread very faſt among the people, owing to our being ſo much crowded, and the ſick lying among

among the fhip's company, not having room in the
brig for a fick birth; we, however, got at laft
clear of it, by perfevering in fmoking the veffel with
fulphur and tobacco every time we went into port:
we found the method of fumigating with nitrous
acid peculiarly ferviceable.

Extract from the Journal of Mr. Alexander Aber-
dour, *Surgeon of his Majesty's Ship Alexander,
between the 24th of July,* 1797, *and the 23d of
February,* 1798.

I would only here obferve, that I have tried the
fumigation with the nitrous acid upon coming from
Gibralter, when we had the fever, and apprehend
that its progrefs was arrefted by it.

Upon fuperintending the bufinefs, I was feized
with a head-ache and flight degree of ftupor;
every one enveloped in the fumes was affected
with coughing.

Extract from the Journal of James Farquhar, *Surgeon of his Majesty's Ship Thefeus, between the
25th of February and the 26th of May,* 1797.

I ordered the fick birth to be regularly fumigated every morning and evening at the time we were
dreffing the ulcers, and found that the nitrous vapour not only purified the air, but in great meafure deftroyed the very fœtid, and almoft intolerable fmell occafioned by the difcharge from the ulcers;

X

ers; the patients themselves likewise found it very refreshing.

I found myself frequently at a lofs for want of a proper veſſel to heat the ſand in; if one or two ſmall iron pots were ordered to be ſent on board with the fumigating materials, they would be found to be very uſeful.

Extract from the Weekly Return of Mr. Thomas Moffatt, *Surgeon of his Majesty's Ship Triumph,* June the 24th, 1798.

Since laſt return, the ulcers have been carefully fumigated morning and evening, and I am happy to add, with conſiderable ſuccefs. They all look clean, and ſome have made a little progreſs in healing already; the fever, which attended ſeveral, in a very great degree ſubſided after three or four days application.

The improvement of the ſmell in the ſick birth is ſenſibly perceived by all.

Extract from the Journal of Mr. Robert Cinnamond, *Surgeon of his Majesty's Ship* Aſſiſtance, *between the 9th of November,* 1797, *and the 5th of June,* 1798.

The only diſeaſes from which infeſtion was to be dreaded were the fluxes, but from a proper attention to cleanlinefs, and a conſtant uſe of the fumigating medicines, I was happy to obſerve there was not one man during theſe laſt ſeven months who
ſuffered

suffered from contagion, their complaints being evidently produced either from expofing themfelves to wet and cold, or drunkennefs. From the fmall experience I have had of the fumigating medicines, I confider them of extreme great fervice.

―――――

Extract from the Weekly Return of Mr. James Rolloff, *Surgeon of his Majesty's Ship Galatea, dated the 5th of August,* 1798.

I find the fumes of the* vitriolic acid of great fervice.

―――――

Extract from the Journal of Mr. Robert Sabine, *Surgeon of his Majesty's Ship Melampus, between the 20th of August,* 1797, *and the 19th of August,* 1798.

The fhip in general has been very healthy, having had no bad fevers on board, which I attribute to fumigating with the vitriolic acid and nitre, (when the weather was fo bad as not to allow the beds to be got up) which always took away the difagreeable fmell there was between her decks, and by airing between her decks with ftoves when they were wet.

Extract

* This gentleman meant the nitrous acid.

Extract of a Letter from Dr. Withering, *of Birmingham, to Dr.* Duncan, *of Edinburgh, publiſhed in the third Volume of The Annals of Medicine.*

It is but ſeldom we ſee much typhus at Birmingham. The uſe of the nitrous vapour, in every inſtance of its adoption, ſtopped the further progreſs of infection, ſo that I am perſuaded we are much obliged to Dr. C. Smyth on this occaſion.

Two LETTERS *addreſſed to Dr.* PERCIVAL, *of Manchester, originally publiſhed (by Order of the* *BOARD *of* HEALTH) *in the* Manchefter *Chronicle, and republiſhed here, as containing a more full Explanation of the Author's Sentiments on the limited Sphere of Contagion, &c. than is to be met with any where elſe.*

London, July 7, 1796.

MY DEAR SIR,

I am now to acknowledge your obliging favour of the 17th of February laſt, to which I did not give an immediate anſwer, being engaged in the experiment on board the Union, which I undertook at the requeſt of the Lords Commiſſioners of the Admiralty

* *Manchefter, Auguſt* 3, 1796.
" That the thanks of this Board be given to Dr. J. Car-
" michael Smyth, for his letters, communicated by Dr.
" Percival, and that they be made public."

Signed by order of the BOARD,

THOS. BELLOTT, *Secretary.*

Admiralty, and the refult of which I was defirous
to communicate to you, as the beft and moft fatisfactory reply I could give to your letter. I defired Mr. Johnfon to fend you a copy of my pamphlet on that fubject, as foon as it was printed.
Although I have been prevented from writing to
you till now, owing to a variety of engagements
and bufinefs, which it is needlefs here to explain, I
can affure you I have not been forgetful of the benevolent undertaking of your Board of Health,
which reflects fo much honour on the gentlemen engaged in it, and to which I fhall be at all times
happy to contribute any affiftance in my power to
give. The very limited fphere of contagion is fo
well afcertained, that I have occafion to fay little
on the fubject. In my book on the jail-fever, I
have mentioned, after many year's experience as
phyfician to an Hofpital, into which more typhus
fevers were admitted in proportion than into any
other, that the moft highly contagious fevers, that
occur in our hofpitals, do not affect the patients in
general, lodged in the fame ward ; for we had no
appropriate fever wards ; nor did I ever fee the
neceffity for fuch, as the communication of infection was in general eafily prevented by the means I
employed. I have alfo mentioned, that there is
no, or very trifling rifk of contagious fevers being
propagated in the open air, ftill lefs from one
room or ward to another ; and that I never knew
contagion propagated by a dead body, even from
the diffection of it, unlefs by inoculation. But, independently of all thefe obfervations, the fumigation with the nitrous acid, if properly employed, not
only certainly deftroys contagion, but improves
greatly the atmofpheric air, by fupplying a quantity of dephlogifticated air, or oxygen gas ; and it
effectually

effectually deſtroys all offenſive ſmell. I alſo, as
you muſt have obſerved, uſe the diluted marine acid
for waſhing the floors, bedſteads, &c. and put marine acid in the pails of water uſed for immerſing
the foul linen, &c. In bed-chambers and private
apartments I generally keep up, where there is a
contagious diſeaſe, a conſtant fumigation; which
can eaſily be done by means of a lamp, over which
is placed a china cup or ſaucer, with oil of vitriol
and nitre, an ounce and a half or two ounces of
each being ſufficient for twenty-four hours. If you
have any queries to put to me, I need hardly aſſure
you, that I ſhall take a pleaſure in anſwering them,
and at all times of convincing you of the regard and
eſteem of, my dear Sir,

Your ſincere friend,

J. CARMICHAEL SMYTH.

Dr. Percival.

London, Auguſt 1, 1796.

MY DEAR SIR,

I am this moment favoured with your letter of
the 30th of July. I can have no objection to your
making any uſe, public or private, of my letter to
you. The accuracy of the facts I will be anſwera-
ble for; but as it was a private communication to
a particular friend, I was little attentive to ſtyle or
manner. Reſpecting the limited ſphere of contagion I ſaid the leſs, as I conſidered it a matter ſo
well aſcertained, and by ſuch a body of evidence,
as required no additional proof. Mankind have
been led into error, on this ſubject, by confound-
ing

ing under the general name of contagious or epidemic, difeafes of very different natures and origin. But of all thofe contagions, that are propagated from one difeafed perfon, or his clothes, to another perfon, the fphere of the deleterious power is in general fo extremely limited, that there have been, and ftill are, fome phyficians, who believe they are only propagated by contact. At Winchefter, during my ftay there, one foldier only, whilft doing duty on the prifoners in the airing-ground, was feized with the diftemper; and very few of the military fuffered, although the guard-room was immediately under one of the prifon-wards, and the fentinels mixed with the prifoners even in the courts and paffages of the king's houfe or prifon. And lately, on board the Union, none of the officers fuffered, and few of the petty-officers; nor would the fhip's company have fuffered fo feverely as they did, could their intercourfe with the nurfes and affiftants in the hofpital have been prevented. But, independently of the limited fphere of contagion, I will venture to enfure even the nurfes and hofpital affiftants, in any fituation, if they will be induced to ufe the proper precautions, and if the hofpital is properly fumigated; the wards fprinkled with diluted marine acid; the dirty linen, &c. immediately immerfed in pails, filled with cold water impregnated with marine acid; the chamber-pots, foiltubs, &c. quickly removed and wafhed with the fame; the bedfteads wafhed every time they are empty with the diluted marine acid; and the bedclothes fumigated with the nitrous vapour. In hofpitals crowded with fick, in fhips, prifons, &c. it is neceffary to fumigate completely every part of the fhip, prifon, &c. twice a day. But in common cafes, and in private practice, fuch means are not

neceffary;

neceffary; and one, two, or three fumigating
lamps, in which a conftant fumigation is kept up,
night and day, fo as to pafs over the beds of the
fick, are perfectly fufficient. In this manner I
have not only ftopped the common contagion in the
hofpital and in private, but I have equally fucceed-
ed, which is of great confequence to be known, in
preventing the fcarlatina anginofa, or putrid fore
throat, from being communicated to the reft of the
family, living under the fame roof. Whether this
will apply to the fmall-pox, I cannot fay from my
own experience; but I have been told by Dr. Rol-
lo, Surgeon to the Artillery, and Mr. Cruick-
fhank, Profeffor Royal of Chemiftry to the Acade-
my, that it deftroys the *miafma* of the fmall-pox;
and that of two quantities of matter, taken for the
purpofe of inoculation, one was expofed to the nitrous vapour, the other not : the perfons inoculated
with the firft were not feized with the difeafe,
whilft the inoculation took the ufual effect, when
performed with the fecond.

I ever remain, with fincere regard,

Your's truly,

J. CARMICHAEL SMYTH.

Dr. Percival.

CONCLUSION.

CONCLUSION.

W HOEVER reads the preceding pages with attention, mult be ltruck with the great conformity of opinion, obfervable amongft the different individuals, in regard to the principal object of our inquiry, viz. the power of the nitrous vapour to deftroy contagion.

But although the fame fentiment univerfally prevails, I cannot help remarking, that whillt the opinion of fome gentlemen is founded on general obfervation alone, the opinion of others is fupported by fuch an accurate detail of facts reduced to the certainty of arithmetical calculation, as carry with them a conviction, little fhort of demonftration itfelf.

The power of the nitrous vapour on contagion once eftablifhed, all its other effects can eafily be underftood and explained; one of the moft obvious of thefe is its deftroying putrid fmell. Although I am far from imagining that putrid fmell and contagion are one and the fame, on the contrary, am convinced that they often exift independent of each other, yet as they are of the fame family, and arife from a common caufe, we may fairly fuppofe from analogy, that what deftroys the one, will prove effectual in deftroying the other. I confefs, that where there is a direct and pofitive proof, as in the

Y   prefent

prefent inftance, reafoning from analogy is of little confequence. The obfervation, however, is in itfelf important, particularly for thofe whofe duty leads them to an attendance on the fick, as the offenfive fmell to which they are expofed, conftitutes not the leaft difagreeable part of fuch an office.

But befides removing the offenfive fmell of hofpitals and prifons, another advantage of the nitrous fumigation is that of rendering the air purer and fitter for the purpofes of animal life; a fact which chemiftry readily explains. From it we learn, that in the decompofition of nitre by the vitriolic acid, a certain proportion of vital air,* or oxygen gas, is let loofe; and phyfiology informs us, that this air, which conftitutes a very interefting part of our atmofphere, is neceffary for the refpiration of animals, at the fame time that it is conftantly confumed by it.

In a former publication I did not hefitate to give a decided opinion, (judging partly from experience, and partly from the fimilarity of putrid contagions) that the nitrous vapour would be found equally an antidote to all,† even to the plague itfelf. I have

now

* " In anfwer to Dr. Smyth's queftion, What is the " proportion of oxygen and nitric acid, difengaged by add- " ing half an ounce of oil of vitriol to half an ounce of nitre?" " I reply that I do not recollect any experiment which has " been publifhed to afcertain the proportion of oxygenous " air thus extricated, though the fact of its extrication is well " known, &c."—*Extract of a Letter from Mr. Keir of Birmingham.*

† This fact becomes the more important, it being now clearly afcertained, that the yellow fever, in America at leaft, is produced by imported contagion. The advantage to a commercial country of being able to counteract all communications of this kind, without fubjecting the merchant to the expenfe and delay of quarantines, is hardly to be calculated.

now the fatisfaction to fee this opinion confirmed, in fo far at leaft, as relates to the dyfentery, a putrid difeafe equally contagious with the jail fever, and in military hofpitals, at leaft, ftill more fatal.

The efficacy, however, of the nitrous vapour, as appears from almoft the whole of the reports, is not confined to the deftroying or preventing the communication of contagion ; its falutary influence is no lefs remarkable on the fick and on thofe recovering from ficknefs ; but on this very important fubject, I could wifh the reader to confult Mr. Paterfon's Table of the Weekly Returns at Forton Hofpital, from which it appears, that during the fhort fpace of fix weeks, in an hofpital containing from 300 to 400 men, there was a difference, from employing the nitrous fumigation, of about 50 lives faved, and about 110 men reftored to a ftate of health fit for active duty ; but if the reader is defirous of forming an accurate judgement of the immediate effect of the nitrous vapour on thofe ill of typhus fever, I would advife him to read with attention, what Mr. M'Grigor and Mr. Hill have written on the fubject.—By Mr. M'Grigor we are told, that fome years back, during the prevalence of a fever fimilar to the one he defcribes, in the fame place, the ifland of Jerfey, the 88th regiment to which he belongs, in the fpace of ten weeks, fuffered a lofs of 40 or 50 men ; whereas during the prefent illnefs, when he employed the nitrous fumigation, of 64 men feized with the fever, he did not lofe a fingle patient. He further remarks, that by ufing conftantly the nitrous vapour, the malignant fymptoms of the difeafe difappeared, and that from a typhus it became a fimple fever.

But of all the advantages to be derived from the ufe of the nitrous vapour, none is more remarkable or likely to be of fuch extenfive application as its effeCt

fect on ulcers, an effect first taken notice of by Mr. Paterfon, and which has been confirmed, upon every fubfequent trial.

That the nitrous vapour, by correcting the malignant and contagious air of hofpitals, which is known to affect* more or lefs, all perfons confined in them, fhould fo far at leaft prove ferviceable to thofe affected with ulcers, and in general to furgical patients; we can readily believe, and indeed it is an induction to which we fhould have been led, reafoning as it called *a priori*. But the nitrous vapour feems to be not only ufeful in this way, it is found of efficacy alfo as a topic or local application; its operation, however, as fuch, muft have its limits; to fuppofe that it will prove a univerfal remedy for all ulcers, is an idea that cannot be entertained for a moment, by any one in the leaft converfant with the animal œconomy, or with the hiftory of difeafes. Thofe perfons who are too precipitate in general conclufions, have commonly fome ground to go back again.

Were I to form a conjecture refpecting the kind of ulcers in which the nitric acid, either in a gazeous or liquid ftate, is likely to be found moft ferviceable; I fhould fay, the floughy or fphacelous, the fcrophulous, and the fcorbutic: but thofe gentlemen, who are profeffionally engaged in the treatment

---

* " Having related the moft diftinguifhing marks of this " fever, I fhall only add, that there are fometimes flight de- " grees of it hardly to be characterized; and which can only " be difcovered in full hofpitals, by obferving the men to " languifh, though the nature of illnefs for which they came " in fhould feem to admit of a fpeedier cure. In fuch cafes " the only fymptoms are flight head-achs, a whitifh tongue, " want of appetite, and other inconfiderable feverifh fymp- " toms." Vid. Pringle on the Jail or Hofpital Fever.

ment of fuch complaints, will look upon this obfervation more as a hint than an opinion.

The preceding effects of the nitrous vapour are
what have been obferved by all or by many; but
there is one which refts as yet on the authority of
Mr. Paterfon alone. He only has made trial of it,
and with fuccefs, in the hooping-cough. His remarks on this fubject, I muft fay, are extremely interefting, and open a wide field for the reflexion and
experience of the practical phyfician.

Having finifhed the few obfervations I had to offer, on the letters and reports which I have now the
honour to lay before the public; the reader I hope
will pardon me if I detain him a few minutes longer,
to make one remark which relates principally to myfelf. It cannot have efcaped his notice any more
than it has done mine, that, as appears by feveral
of the letters, there are prejudices entertained
againft the nitrous fumigation by many furgeons of
the navy. At this I am by no means furprifed; we
are all children of habit, and unwilling to relinquifh
opinions which we entertained in early life: the introduction, however, of the nitrous fumigation into
the navy, has been oppofed not by prejudice only,
but by arguments drawn from chemiftry. It would
be no very difficult tafk for me to point out the fallacy of fuch chemical reafoning, but to endeavour
to refute by argument what is directly contrary to
experience and obfervation, would be an abufe of
time, and an infult to the public judgment. The only
anfwer then that I fhall give to fuch philofophers, is
to addrefs them in the words of an author, whofe
opinion muft be confidered of high authority on fuch
a fubject, as he was not only a phyfician of character, but certainly one of the firft chemifts in Europe;
the circumftances and occafions were perfectly fimilar.

Comme

" Comme il* n'a certainement eu en vue que le
" bien de l'humanité, il me permettra quelques re-
" flexions qui ont bien pu échapper au favant chi-
" mifte, mais que ne pouvoient manquer de frapper
" un médicin, qui quoique amateur zélé de la chimie
" & convaincu des avantages qu'elles peut procurer
" à l'art de guérir, a été trop fouvent témoin des
" erreurs que cette fcience a portées dans la médi-
" cine, pour n'etre pas toujours en garde contre
" elle ; d'autant plus même, que fes raifonnemens
" font plus feduifans, & fes expériences en appar-
" ence plus concluantes."

Mémoire par Monfieur Bucquet,

    Profeffeur de Chimie, Cenfeur Royal

    De L'Academie Royal des Sciences, &c. &c.

* The perfon alluded to was Le Sage, who oppofed chemical reafoning to experience and obfervation.

FINIS.

www.ingramcontent.com/pod-product-compliance
Lightning Source LLC
Chambersburg PA
CBHW021806190326
41518CB00007B/472